COMPETITIVE MANUFACTURING THROUGH INFORMATION TECHNOLOGY

COMPETITIVE MANUFACTURING THROUGH INFORMATION TECHNOLOGY

The Executive Challenge

JOHN STARK

VNR VAN NOSTRAND REINHOLD
New York

Library of Congress Catalog Card Number 89-70530
ISBN 0-442-23932-7

Printed in the United States of America

Van Nostrand Reinhold
115 Fifth Avenue
New York, New York 10003

Van Nostrand Reinhold International Company Limited
11 New Fetter Lane
London EC4P 4EE, England

Van Nostrand Reinhold
480 La Trobe Street
Melbourne, Victoria 3000, Australia

Nelson Canada
1120 Birchmount Road
Scarborough, Ontario M1K 5G4, Canada

16 15 14 13 12 11 10 9 8 7 6 5 4 3 2 1

Library of Congress Cataloging-in-Publication Data

Stark, John, 1948-
Competitive manufacturing through information technology: the executive challenge / by John Stark.
p. cm.
Includes bibliographical references.
ISBN 0-442-23932-7
1. Business—Data processing. 2. Management information systems.
3. Information technology 4. Chief executive officers. I. Title.
HF5548.2.S7735 1990
658.4'038—dc20

89-70530
CIP

To
Alicja, Jasna
Chamichou
and Lawrence

CONTENTS

PREFACE

This book is addressed to the top management teams of manufacturing companies. It describes how Information Technology (IT) and communications can be used successfully within their companies. In particular, it shows how IT can be used to gain competitive advantage.

The book is in four chapters. The first chapter describes why Information Technology is such an important tool for manufacturing companies, and why top management has to play a leading role in setting IT strategy. The second chapter explains the actions that have to be taken by top management if IT is to be used to support the business strategy and win a competitive advantage. Then, examples are given of the way that companies from many industrial sectors are using Information Technology to gain competitive advantage in their everyday business activities. The last chapter of the book looks at some of the many ways in which manufacturing companies can use IT for competitive advantage.

JOHN STARK

Robert E. Umbaugh: I read in the *Harvard Business Review* about companies saying they make strategic use of information—that they use information as a strategic weapon.

Peter F. Drucker: I don't know what that means.

Robert E. Umbaugh: I think what these companies are saying is that they are taking advantage of their enhanced systems.

Peter F. Drucker: Well, some people are much further ahead than others, but only a very few are really far ahead, and very many are very far behind. There still are people who don't know how to do inventory on computers—more than you are willing to accept. They haven't learned for instance, that when you run a market test in Denver on a new toothpaste, you better get results every day and not every month. And get them in a form that shows [these results] by store and, if possible, whether this is the first or second purchase. Which, as you know, we wanted all along and never could get. Now we can get it. You use the information for a strategic purpose because it enables you to make decisions about whether or not to put more promotion money on the product. But if you can get this information, you can spend promotion dollars far more effectively than we can now.

From: *Handbook of MIS Management,* 2nd edition. Robert E. Umbaugh, Auerbach, New York, 1988.

COMPETITIVE MANUFACTURING THROUGH INFORMATION TECHNOLOGY

1

FROM EDP TO IT

1.1 SETTING THE SCENE

The objective of this book is to provide the top management team of a manufacturing company the information needed to use Information Technology (IT) and communications to gain a competitive advantage.

This book sets out to help the top management team get back into the driving seat on IT issues. It gives top managers the information they need to blend their knowledge and experience to provide competitive advantage for their companies.

Recent surveys show that although top managers apparently recognize the importance of IT, most of them are poorly informed to manage it. Typical results prove that less than one-third of top managers feel that they are using IT successfully. They feel they need it and could use it to gain a competitive advantage, yet do not understand enough to be really sure or to back their hunches with hard cash. Many top managers have received no formal IT training and have not been in everyday contact with it. They have often hoped that it was a technical issue to be left to the specialists. They do not know how to set an IT strategy that relates to overall business strategy.

The allocation of resources is a major top management activity. IT spending often accounts for 2.5% of revenues, and 50% of capital expenditure may be IT related. Although top managers know that they should be controlling this spend, many realize that they are not sufficiently aware of where it is going and certainly do not control it. They will not know how to measure the results of the IT investment or even which measures to use.

Top managers realize that marketing, engineering and manufacturing processes are changing fast under the influence of IT, and they know that management processes and organizational structures must change correspondingly. They read about other companies using IT to reduce product development time, to reduce batch sizes, to increase quality, and to improve the productivity of their sales force. When they look at the way their own company is using IT, they see nothing so likely to help the company gain or maintain a competitive advantage. They feel instinctively that they are missing out on something—but they do not know what to do about it.

They realize that the way a company organizes for IT, and uses it, is rapidly becoming a key issue. They know that this is important, and that they should be leading the company's response, or at least be involved—but they don't know how.

Often they do not really know why people talk about IT and whether or not IT is the same thing as DP, EDP or MIS.

To make things clear from the start, DP, EDP, and MIS are acronyms from the past. In this book they will be referred to as the traditional MIS—EDP environment. The future environment will be referred to as the IT environment. Table 1.1 shows some of the differences between the two environments.

IT and communications are strategic weapons that enable the company to compete successfully in a changing world full of competitors, suppliers, customers, and other players that affect the business environment. MIS—EDP was a tool that focused on the internal working of the company, in particular the Finance and Administration Department (F&A). The objective of IT is to help the company gain competitive advantage. The objectives of MIS—EDP were to make the workings of F&A more efficient and to produce information for management.

The key technology for MIS—EDP was the mainframe computer, and the key people were the MIS team. In the IT environment, the key technologies are the computer programs, or applications, on the computers at the work-place of everyone in the company and the communications system that links them together and to the outside world. The key people are the users of these applications, not the IT support team.

The subjects of the book are management and organizational, not technical, issues. The book is not about the hardware and software details of IT. There are already thousands of books full of details about computers, networks, bits, bytes, bauds, workstations, relational data bases, X.25, X.400, and so on. The book does not attempt to address the technical details of these subjects. It does not aim to list the advantages of every computer-based application or to examine, for example, the merits of electronic mail relative to voice communications. No technological expertise is needed to understand this book. Successful use of IT at the company level is a management issue—not a technical one.

Table 1.1. Differences between the traditional MIS-EDP environment and the future IT environment.

MIS—EDP Environment	IT Environment
Little or no top management involvement	Top management involvement essential
Computers and communications are not strategic tools	IT is a strategic weapon
The key technology is the mainframe computer in the computer center.	The key technology is the application in the computer at the work-place of everyone in the company
The key people are the MIS Department (i.e., support staff)	The key people are the "end-users" of computers, not the support staff.
The product of MIS-EDP is the program	The product of IT is business use of programs
Computers only used for internal company application	IT helps links the company to the rest of the world
Departmental use of computers. Use focused on F&A Department	IT plays an important role in all departments including engineering, manufacturing and marketing
Computers used to reduce direct costs and provide information to management	IT focused on gaining competitive advantage
Investment justification for computers based on dollar savings in direct costs	IT investment justification based on increased revenues from increased market share and reduced time to market as well as reductions in direct costs

The book takes a distinctly partial point of view. It does not offer an academic, neutral review of the ways in which manufacturing companies might make use of IT. It does not debate whether IT should or should not be used to gain competitive advantage. It assumes that it should and shows how it can be done. The book is a practical guide for busy people. Given its aim, it is as short as possible. Its pages are not full of references to learned business reviews and other sources that most managers neither have to hand nor would bother to refer to. Neither is it expected that busy managers reading this book would want to be side-tracked by references to details that are not focused on the objective of the book.

The book communicates a message— "You, top manager, need to understand why IT is so important for your company. Once you have understood this, and how to apply it to your company, you are on the road to success."

The book champions IT as an important weapon in the business armory. This may lead some top managers to put the book down in disgust. They may see this as just another technological crusade—another CAD (Computer Aided Design), JIT (Just-In-Time), or TQC (Total Quality Control): it is not. IT strategy is a key part of the business strategy of a manufacturing company and on average has a higher place in the hierarchy of manufacturing acronyms than CAD, JIT, or TQC. IT affects all functions, not just Engineering or Production. Information is everywhere in the company; Information Technology is a core activity. Top managers can "hive off" manufacturing, put it "offshore," choose to become a "hollow" company—but they cannot hive off all information-related activities. They may be able to sub-contract activities, such as operating the computer, since this is no longer a strategic activity. However, most of the information-related activities are much too closely intertwined with the strategic activities of the company to be divested. They must be kept and developed, and a lot of top management activity will need to be spent on them.

The book does not claim that IT is always going to be the most important activity in a company, but there are very few companies in which it will not be one of the most important activities. It will not be claimed that IT will solve all the problems of the company, nor that IT can be successful without major investment in time and people. IT alone cannot solve management and organizational problems in the company.

Ideally, the first person in the company to read the book should be the CEO. Because IT is strategic and cross-functional, the CEO is a key individual. If the CEO does not like the book, then it should immediately be put in the trash can. (Few managers will want to unnecessarily oppose the CEO.) If the CEO likes it, then the rest of the top management team should be encouraged to read it. (The book is addressed to the top management team.) It is very unlikely that one manager alone, even the CEO, can, without support from colleagues, put the company on the right road. Information is everywhere in the company—marketing, product development, research, manufacturing, finance, and distribution. The best IT solution will involve developments and changes in several of these areas, and require support from several managers. It is best, from the start, to gain a common understanding among the top management team. These individuals are responsible for business strategy, make major decisions, and set company culture. It is at their level that the major decisions concerning IT have to be taken. A common background and a common approach to IT will provide the top management team with a good basis for using IT to gain competitive advantage.

Once the book has been assimilated by top management, it will be useful to distribute it to the middle management team. This will help them understand the approach that the top management team is following and to begin taking account of it in their everyday activities. Initially there may be little to be gained by distributing this book to the company's analysts and programmers. They do the job that management tells them to do—and they do it as well as they can. There is not a lot more they can do, if the message from the top is not the right one. How can they write a program that will provide competitive advantage? How many of the company's programmers know enough about the 5-year business strategy to be able to understand what competitive advantage might involve? What chance can someone in Engineering have of building a competitive advantage if people in Marketing and Manufacturing are working along completely different lines? The only people in the company who can really start building a competitive advantage through the use of IT are the ones who know the business mission and goals and who have company-wide authority, i.e., top management.

However, once top management has set the strategy, and the implementation plans have been agreed, the company's analysts and programmers will be involved in implementing systems that will give the company a competitive advantage. They will only be able to do this successfully if they have been introduced to the new concept as soon as possible. Similarly users of computers in other departments need to be made aware of what is happening. A culture of "IT for competitive advantage" needs to be built up, and this implies that everyone needs to talk the same language and understand the common objectives.

1.2 NOT ANOTHER ACRONYM

Acronyms can be helpful: they condense into a few letters what would normally take a few paragraphs to explain. People often interpret acronyms differently, and the details of their interpretations differ greatly. As an example, consider the acronym CIM (Computer Integrated Manufacturing). What is the meaning of the phrase "We have CIM in our business." Does it mean that there is a FMS (Flexible Manufacturing System) or a CAD system in the company? Does it mean that the CAD system is linked to the MRP 2 (Manufacturing Resource Planning) system? Does it mean that shop floor computers are linked to MRP 2? Perhaps it means that the company has completely eliminated the use of paper, and all information can only be found on the computer. This is a typical example showing how acronyms are not necessarily useful in transmitting detailed information.

Acronyms can be helpful to authors though. It would be so easy to say that the subject of this book was "XYZ," but unfortunately it is not that easy. One of the subjects of the book is Information Technology, but it is not the "Technology" itself that is of interest—it is more the management and the organization associated with it. A title such as "Information Technology—Management and Organization" would not be complete either because the book is also about the way IT strategy needs to be linked to the business strategy, and the way that competitors can be beaten. Perhaps one could use "Information Technology—Competitors, Business Strategy, Management and Organization," but even this does not mention the customers and the suppliers. Unfortunately, there is no existing acronym that covers all of the issues, and if one was invented it would run to about fifteen letters. The lack of a specific acronym makes the subject more difficult for unwilling readers who will have to try to understand the book rather than just learn an acronym: to ease their pain, this book has been kept relatively short.

Can it be true that there is not a suitable acronym? How about DP (Data Processing) or its cousins EDP (Electronic Data Processing) and ADP (Automatic Data Processing)? Not really. These acronyms describe the use of a computer to perform a series of operations on data—sorting, classifying, calculating, storing, and so on. They conjure up images of large computers equipped with rows of flashing lights, cardboard boxes overflowing with discarded listings, and grey tin boxes protecting dog-eared decks of punched cards. They don't say much about telecommunications, business strategy, or the way the company should be organized.

Would MIS (Management Information System) be better? From the sound of it, a MIS should give all levels of management the information they need to perform their work. It should be a system that provides management with information. One can interpret "management" in any number of ways, the same is true of "information," and even more so of "system." However, taken word by word, it reads like a system that provides information for management. No sign of the business environment or strategy, no sign of management actually doing something. It is, like DP, focused on technology—computers and systems that have their own minds and do things to data and information, and then perhaps, to justify their existence, provide information to management. Additionally, MIS is an emotive acronym. In many manufacturing companies, it was the name given to a department that lost its way. The MIS department concentrated on providing a limited service to a few managers and a few users of computers in the Finance and Administration activity of the company. In doing so, it ignored and alienated many potential users of computers in other functions, such as engineering, marketing, and manufacturing. Clearly, MIS is not a suitable acronym for the subject of this book.

Information Systems (IS) could be a good acronym, with the "systems" implications of an organization, and things working together. It is open to the same criticisms as MIS, and like MIS does not convey the competitive and business issues that the book addresses.

In many manufacturing companies, "Information Systems" also has negative connections. MIS was a bad enough title, with the MIS people perceived as never understanding the real needs of people in a manufacturing company. With a title such as "Information Systems," the computer and communications function would appear to have drifted even further from reality—even the word "Management" disappeared from the title. It no longer dealt with tangible issues such as management, people, products, and cabling but floated somewhere up in the clouds with the doubly intangible Information (which is everything and everywhere) and systems (which can take on an infinite variety of meanings).

In companies that have one, the Information Systems Department is typically an inward-looking, technology-dominated, hierarchical, standardized and centralized closed shop. These adjectives, which describe the culture of the beast, can be applied equally well to its views of the company, of the business environment, and of computer, data, and communication architectures. They also apply to its internal organization and to its relationship with other users of IT in the company. The members of the Information Systems Department are typically IS wizards (or freaks) who know the finest details of computer systems and programs but do not understand the business issues behind them. They speak in specialized jargon and adopt unusual behavior that leaves them far from credible with users of the systems they try to develop.

Information Technology (IT). Clearly, this acronym looks as if it should be a loser. It sounds like computer hardware—computers, workstations, disks, and tapes. It is not immediately obvious if the acronym includes software or communications networks. "Technology" is defined in different ways, but generally it is linked to the study of the tools, techniques, and procedures associated with an industry: that is a fairly wide range. Under the IT label, one can classify computer hardware and software, programming, analysis, communications, and use of computers. The management and organizational issues could be classed under the headings of techniques and procedures, but there is not a specific reference to competitive issues.

It seems as if all the acronyms commonly in use are related to information, computers, and systems but not to their use in the business environment: that in itself tells a story. Using IT for competitive advantage means going into uncharted, "un-acronymed" territory. However, to avoid inventing a new acronym, Information Technology will be referred to as the acronym of the future. To differentiate between the way things were done in the past, the acronym MIS—EDP will be used to describe the bad old days when IT was never used to gain competitive advantage.

1.3 MIS—EDP AS IT WAS

Before moving on to describe how IT will be used in the future to gain competitive advantage, it will be useful to look back at how Information Technology has developed.

There appear to have been some six to eight types of activity, and these have occurred over four technological eras (see Table 1.2). These periods have overlapped slightly. In the first, the mainframe computer was dominant (from the early 1950s to the late 1970s). In the second, the minicomputer became important (from the early 1970s to the early 1980s). The third period saw the arrival of the microcomputer (from the early 1980s to the late 1980s). The fourth period, that of open systems and networked cooperative computer resources, began in the late 1980s.

The very first activity was to automate routine operations. Examples of this type of activity can be found all over the company. In the Finance and Administration function, payroll, billing, and accounting were automated. In Marketing and Sales, bulk mailshotting was eased by putting address lists onto the computer. Production Planning spanned major new applications of stock control and MRP (Material Requirements Planning). On the Shop Floor the control of machines was automated by the introduction of machine controllers and the corresponding machine programming languages. Communications was seen as a necessary evil of underfloor cabling.

Table 1.2. MIS-EDP and IT activities.

Period	Activities
Mainframe Computer	—Automate existing processes —Improve decision making and management control —Improve operational efficiency
Minicomputer	—Automate existing processes —Integrate
Microcomputer	—Automate existing processes —Provide additional local support —Implement communications network —Implement information flow applications —Integrate
Networked cooperative computer	—Develop communications networks —Integrate —Use IT for competitive advantage

Looking back one can ask if any of the systems implemented at this time were supposed to provide competitive advantage. In most cases, it appears to have been more a case of automating whatever could be automated. In many cases, completely unnecessary and useless processes were automated. One of the worst examples of this is in the area of stock control where, rather than reorganize to minimize stocks, many companies developed the most complicated and "sophisticated" procedures and systems to handle completely superfluous stocks.

Once the initial urge to automate was starting to generate its own momentum, the second type of activity was launched. These were the Management Information Systems (MIS), designed to provide information to support managers taking decisions. One is tempted to ask the questions—What management? What Information? What for? In principle, these systems should be very useful, but do top managers really need masses of out-of-date information? Rather than the system automatically serving up massive volumes of data from automated operational systems, should it not be the other way around—with the top manager requesting ten or twenty key items of information that give more control of the business? These systems have been built on the belief that "managers want whatever the system gives them," rather than "the system gives the managers whatever they require." Top managers using these systems found themselves deluged by a flood of data, with very little real information available. It was difficult to make decisions when computers from different departments produced different versions of the same facts. Many MIS systems produced vast amounts of data showing ludicrously high stocks and long product delivery times—yet few managers did anything about it. What is the value of data to someone who does not know how to interpret it?

During this phase, the cost of communications began to mount, but compared to the cost of mainframe computers it was generally a minor cost that was allowed to grow unchecked.

With the introduction of minicomputers and then superminicomputers, more activities could be tried. Typically though, the first response was, again, to automate manual activities. For example in the Engineering Department, CAD (Computer Aided Drafting) systems were introduced to automate the drafting process. With the passing of time, progress was made and it was possible to go beyond pure automation of the manual task, and offer new possibilities. For example, CAD/CAM systems helped engineers carry out tasks that were not possible on a drawing-board. During the minicomputer period, it became possible for functions such as Engineering to install their own computers, rather than rely on the centralized EDP mainframe computer. Each department spent months, or years, in defining its particular needs and then in benchmarking systems from several vendors. The analysis rarely took account of interactions with other departments, looked at the past rather than the future, and considered local, tactical issues rather than issues of strategic importance to the company. Although it appears that in many systems, the software costs twice as much as the hardware, and associated organizational issues cost three times as much as the hardware, the benchmarks generally concentrated most on the hardware and least on the organizational issues. The cost justification of such systems often had little value apart from satisfying top management's requirements for paper figures on Return on Investment (ROI). The financial analysts rarely understood the overall impact of new systems, and the potential users of the systems would go to any lengths to get over the hurdle rate set by management.

Up to this point in time, most people in the company had been doing pretty much as they liked when looking for computers. Each department, or group, had bought its own computer and systems to meet its own requirements. Many major corporations found themselves unable to do business without Information Technology but saddled with a wide range of computers that did not communicate effectively together. Trying to get information from a program on one computer to a program on another computer was extremely difficult. The next activity that had to be launched—"integration" was aimed at connecting the "islands of automation" together. Whether or not this was necessary from the point of view of the business strategy was rarely considered.

At about the same time as the integration issue became apparent, the first microcomputers became available. Along with Personal Computers and Engineering Workstations, they offered new possibilities to improve operations. It became possible to automate more manual activities and for the first time to provide specific local computer support to individuals. Word processors replaced typewriters. Sales forces were equipped with portable computers, marketers developed their own databases and typists became word processors.

Before the explosive growth in the use of microcomputers, telecommunications had been a separate area, dominated by engineers adept at splicing cables and interpreting signals on oscilloscopes. As more and more people started to use their own, rather than a centralized computer, the need for communications networks to transfer information between users became apparent.

The rapid development of communication networks provided the opportunity for another IT activity—communicating information throughout the company. Corporate mail-boxes and electronic mail became familiar buzzwords. Little attention was paid to investigating whether moving information quickly around the company without a massive investment in training and modifying work practices would actually provide a competitive edge.

Finally, the latest activity appeared—"using IT and communications to gain competitive advantage." IT at last had top management attention. Companies understood they could use it to create close ties with suppliers and clients. They could use it, as examples later in the book show, to help differentiate themselves from the competition through lower costs, higher quality, and faster reaction to market requirements. It could be used to specifically address niche markets. These are strategic issues and implied that the decision to invest in IT ranked alongside, for example, major product investment decisions concerning a new product. As a result, they required significant top management involvement, going far beyond signing off on the MIS Manager's budget plan.

Looking back, it is surprising how little the MIS—EDP activities were related to the real problems of increasing sales, developing new products, and finding new clients. Instead, the main objectives have been, regardless of the real business requirements, to reduce costs through automation and to use technology wherever technology was usable.

1.4 WHAT WAS WRONG WITH MIS—EDP

For IT to be successful, it has to be closely linked to the business strategy. Too often in the past, this has not been the case. MIS and EDP have been treated almost as if they were not part of the business, as if they were not manageable. They seem to have fallen into the hands of those who used and supported them in the early stages (generally Finance and Administration and the EDP Department). These people soon found that they had a better understanding of computers than the rest of the company, which they soon considered as being too ignorant to communicate with. The idea that computers could be used at the heart of the activities of most manufacturing companies, on the shop floor, was contemptuously dismissed. Their use in Engineering Departments was rarely supported by the EDP Department, and engineers, rather than engineering the company's products, had to waste their time learning how to select, use, manage, and maintain computer systems.

No wonder that, with such a history, most top managers do not consider IT as a major part of their business strategy or dare to use it to try to gain a competitive edge. They still see IT as hideously expensive, unmanageable, and run by "techies" who have no understanding of the business and hide behind their technical jargon. They see their staff taking three years to produce an IT plan that will be out-of-date six months later. They hear of hardware costs halving every two years, yet see IT budgets spiralling upwards. They see computers being upgraded, and programs rewritten and modified to such an extent that maintenance costs amount to 80% of the budget. If they ask about a potentially useful application, they may be told to wait two years or more until the backlog is cleared, or to provide money for more staff. They see over 40% of projects being delivered late and another 40% over budget. They see each department fighting to have its own computers—empire-building at its worst, since the less a department's computers communicate with anyone else, the more secure the department will feel. They see the EDP department supporting the parts of the business it chooses to support, while diverting resources from others that are often much more important.

Is it reasonable that top management be expected to buy off on a plan from the EDP Department that provides for the widespread implementation of word-processors, work-stations, and electronic mail at a time when the company's products are late to market, slipping behind the competition, delivered late to clients, and not cost-effective? Is electronic mail going to change the situation? Do customers prefer to have their orders delivered on time or to receive a beautifully word-processed letter of apology for late delivery?

Yet, who is really to blame for this state of affairs. Who is supposed to be running the company? Who appears to have abdicated that responsibility in respect of Information Technology? Who has delegated IT too far down the ladder? There is only one answer—top management.

The solution is clear. Information Technology must be treated as an important part of the business, and the development of the IT strategy seen as an integral part of the business strategy. Top management must start taking the time to understand what IT can do for the business. It is not enough to rubber-stamp the plans produced by the EDP department, allowing its annual budget to rise by the rate of inflation. It is not enough to allow EDP only to be used to try to streamline F&A direct costs, while the strategy issues are ignored. Top management must participate in the development of an IT strategy that clearly tells the IT people what they have to do—the IT people shouldn't be telling top management what to do. After all, who runs the business?

Top management has to take control of IT. It has to use it to gain competitive advantage. IT has to be used as a weapon in tomorrow's outside world (competitors, clients, suppliers) not a weapon to fight yesterday's internal battles (reports on shop floor absenteeism). In the past, MIS—EDP has so often appeared to have no objective other than the perpetuation of the EDP department. Top management has to set the IT objectives—and most probably these will be mainly directed outwards to the real world where products are sold, and not inwards to the support functions, where one is limited to controlling the situation through rationalization, efficiency, and order.

It is easy to criticize top management for letting the current situation develop. It is also easy to see how the situation arose. Top management cannot be expected to get involved in every technical issue in the company, and EDP was a very technical issue. Top management has expected the technical EDP specialists to manage IT properly. Unfortunately, EDP specialists are not always interested in the business issues and in time their dialogue with top management has faded to an annual demand for more dollars. The EDP specialists did what they saw to be their best. They were asked to do all sorts of things, but they weren't given the resources to do them all. They often reported to a manager in the F&A organization who had little understanding of either IT or the business issues, and was under orders to keep costs down.

This manager was rarely the most innovative person in the company, and a request from the EDP Manager to fund the fifth redesign of the payroll system in eight years was more likely to receive approval than the first request to use IT for competitive advantage. When EDP work was questioned, the EDP Manager could justifiably point to rising productivity as measured in lines of code produced per programmer.

When potential users of IT, exasperated by the lack of service from the EDP Department tried to install systems in their own departments, the EDP Manager often fought to prevent this, claiming that the advantages of centralized computing far outweighed those of decentralized computing. The EDP department had to stay alive, and if this end would be achieved by buying a new mainframe computer and installing word-processors and electronic mail but not by putting local area networks and scheduling systems on the shop floor, then the choice was clear.

So in time, the MIS—EDP function, ignored in the business strategy, became more and more a support function for the management, financial and administrative activities of the company. It became less and less involved in helping marketing, engineering, sales, and manufacturing to produce quality products quickly and to increase market share. It came to be seen internally as lacking direction, expensive, inflexible, and offering a low-quality, behind-schedule service. Externally, it just wasn't seen at all: it had ignored the competitive world.

As a result, top management had to act. The next time that the business strategy was discussed, top management would have to make sure that an IT strategy, supporting the rest of the business strategy, was clearly defined.

1.5 THE NEED FOR A MODEL

A model is a representation, preferably simpler, of something else. Figure 1.1 is a typical model of a company's organization. It shows who runs the company and how the major functions are departmentalized. Clearly though, it does not help in showing how the company can gain competitive advantage. It is not a model designed for that purpose. Figure 1.2*a* is a typical model of a company and its environment. Figure 1.2*b* is a similar representation including some of the potential internal sources of competitive advantage. These models are more useful in helping to understand how a company should use IT to gain competitive advantage. They are very much oriented towards the business environment, not towards an EDP or even a departmental view of the world. On the other hand, Figure 1.3 and Figure 1.4 are models that are more oriented towards EDP. The data flow model shown in Figure 1.4 depicts how information flows between a manufacturing company and the external environment. Figure 1.5 shows an alternative representation of data flows that includes internal data flows. There are numerous models of this type available, but unfortunately none of them are very helpful when it comes to understanding how to use IT to gain competitive advantage. In this respect, more business-oriented models, such as Figure 1.5 are the most useful. To understand how IT can be used to gain competitive advantage, it is first necessary to understand where in the business and company environment a company should be looking to gain competitive advantage.

Models are simple representations of more complex entities. Because companies are very complex entities, models can be useful in describing them. Similarly, because major IT systems are complex, models can be useful in describing them. As with an acronym though, there is a danger that a model will be understood differently by different people and that

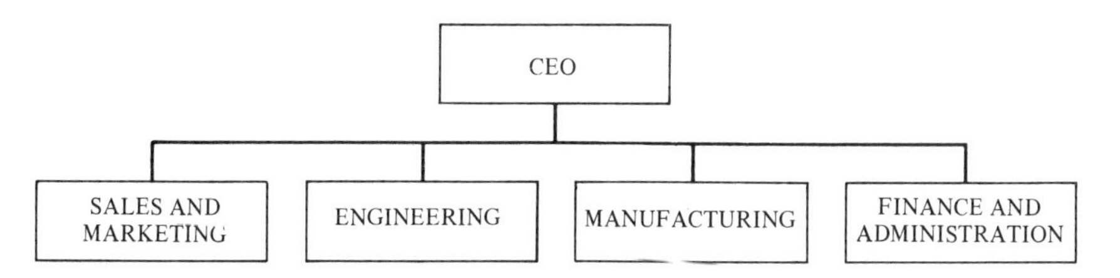

Figure 1.1. The company organization.

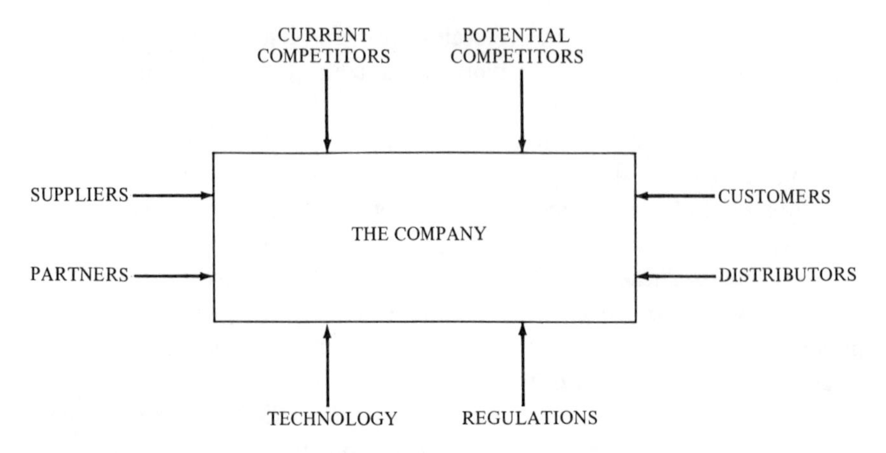

Figure 1.2a. The company and its environment.

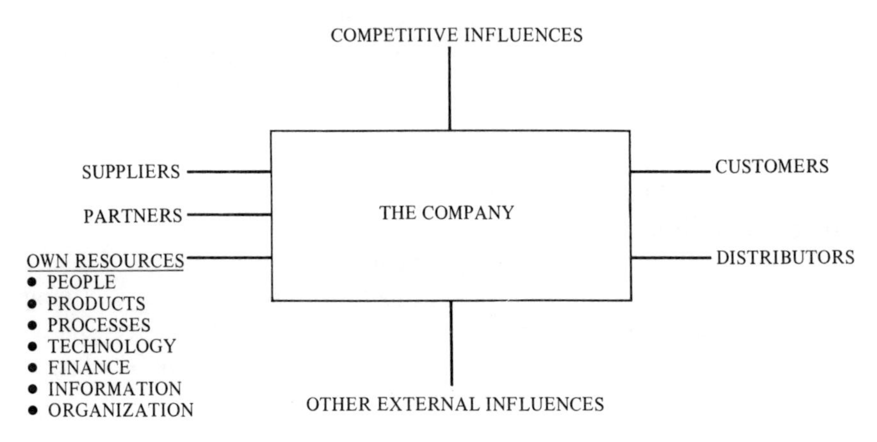

Figure 1.2b. Potential areas of competitive advantage.

in the search for simplicity, the existence of a model will focus attention on particular characteristics of the modelled entity, and other characteristics will be ignored.

As an example of this, consider again the typical model of a company's organization shown in Figure 1.1. It shows the four major functions of the manufacturing company reporting to the CEO. Figure 1.6 is a more detailed

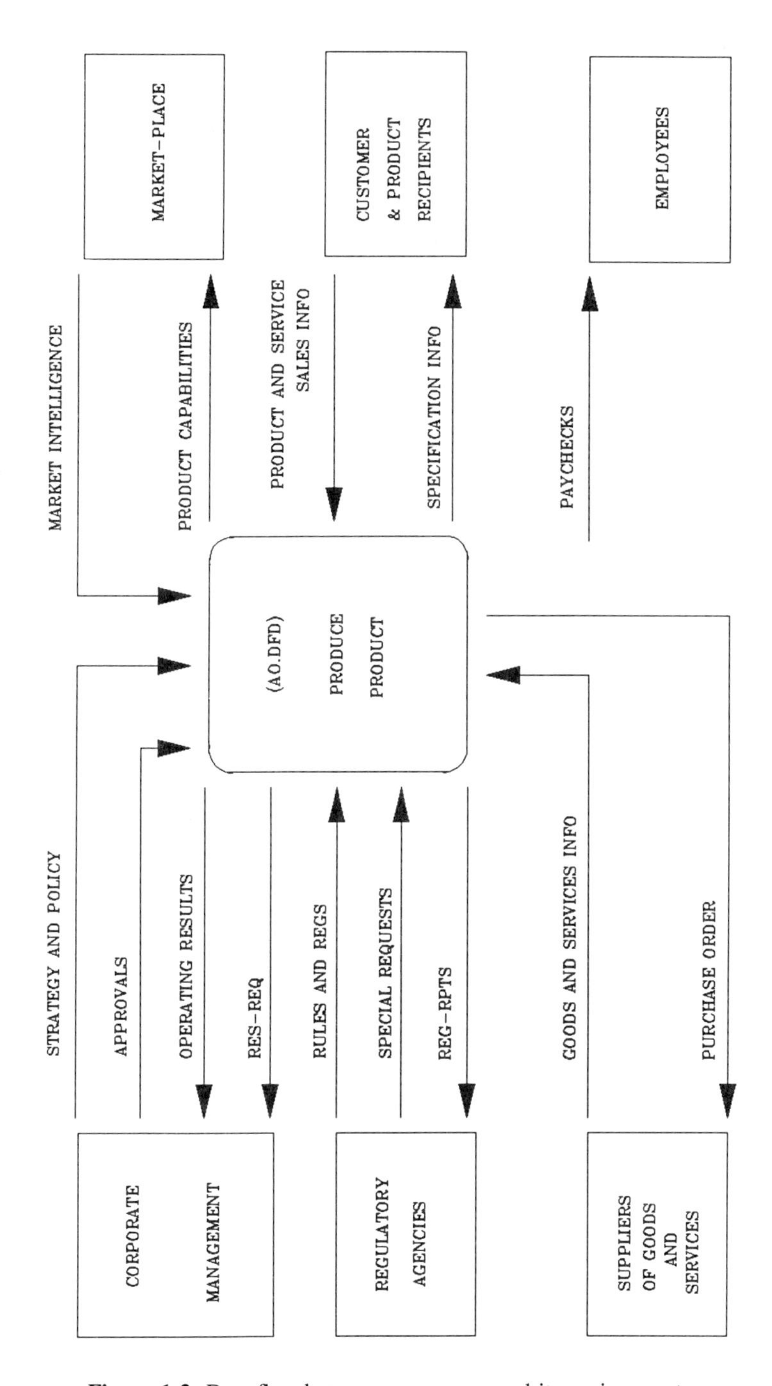

Figure 1.3. Data flow between a company and its environment.

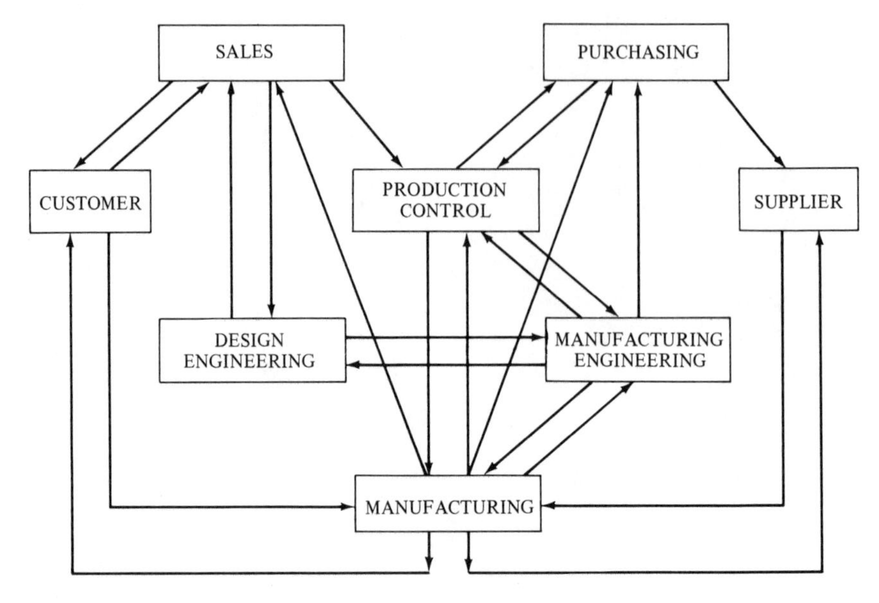

Figure 1.4. Data flows.

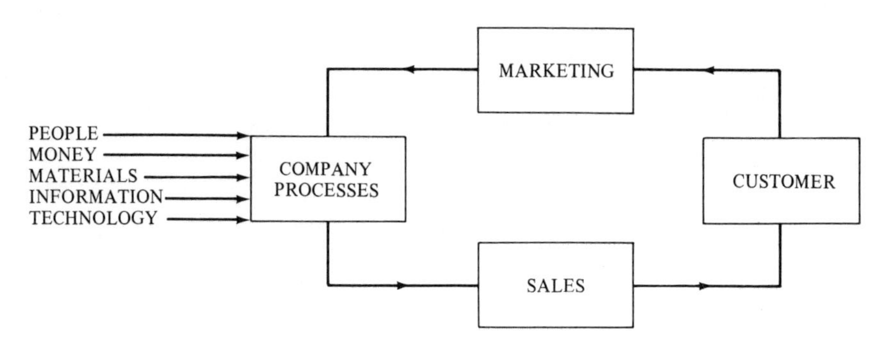

Figure 1.5. Starting to identify competitive advantage.

version of the company's organizational structure. In this company, it had been decided that there would only be four people reporting directly to the CEO. The model shows how the other managers report to these four people. A model such as this is a good starting point for functional managers who are interested in empire-building, who want to have many people reporting

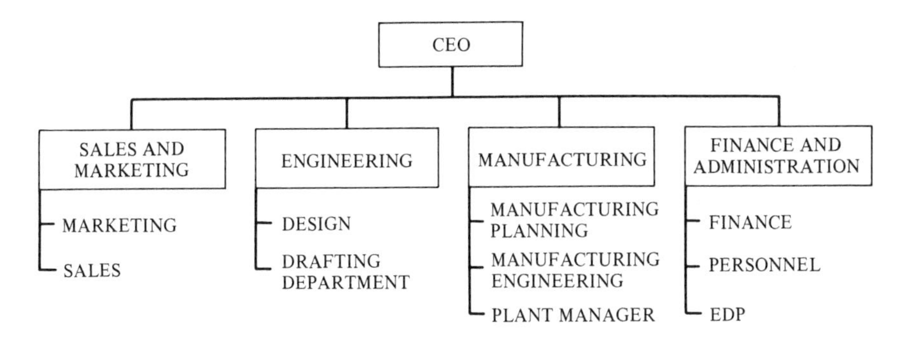

Figure 1.6. More detailed company structure.

to them, and hope to monopolize company resources for their own benefit. Unless great care is taken, such a model will reduce communication between design engineers and manufacturing engineers. They report to different people, and these people will probably not be working toward the same goals. Similarly, the EDP Manager reports to the F&A Manager, who will no doubt assume that the EDP Department only needs to provide service to F&A. Unless great care is taken with models, they can easily be misused and their initial objectives forgotten.

A more positive use of models is in business modelling where the aim is to try to understand and clearly illustrate how the company works. This is a useful exercise both from a specific IT point of view and from the broader viewpoint of increasing overall understanding of company operations. In the IT context, it is useful because it can show how data is used, stored and communicated throughout the company, and how the different parts of the company work with each other and with the outside world. On the more general level, it is a useful exercise in which everyone can participate and contribute to developing an agreed description and understanding of the way the company works. Business modeling can provide a concise picture of the activities of the company and the relationships between both the different parts of the company and the external environment. Figure 1.7 and Figure 1.8 show how, starting from Figure 1.3, it is possible to gradually increase the level of detail in the model.

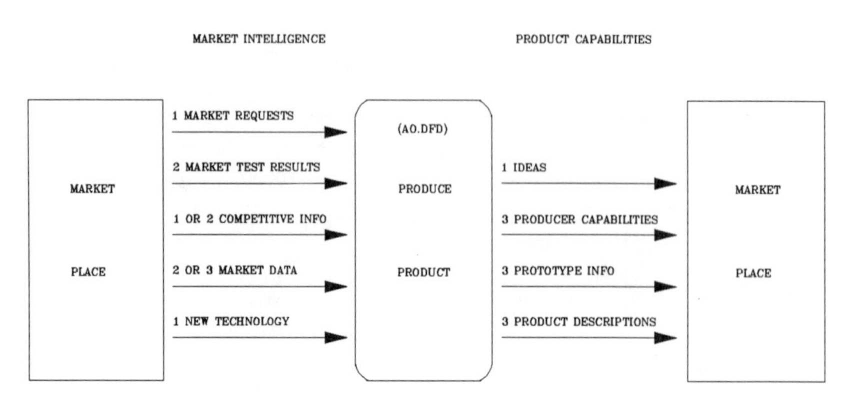

Figure 1.7. More details on information flows with the marketplace.

1.6 THE FUTURE MANUFACTURING MARKET

By looking back at the way EDP, MIS, and IT have been used, it is possible to understand how the current state of affairs was reached. Similarly, if the future environment of the manufacturing company can be foreseen, it should then become possible to understand which IT related actions should now be taken to prepare the company for the future. As a first step, it is useful to look at how the manufacturing market has evolved and to try to understand how it will change in the future. This will help in understanding the future role of IT in the business.

Looking back, the 1960s appear as a golden decade in which manufacturing companies ruled the world, grew steadily in their domestic markets and made tentative forays into international business. There was steady world growth, and relatively homogeneous mass markets for consumer products. The 1970s, with the oil shock and spiralling inflation, caused distress for many manufacturing companies. Worse was to come in the 1980s, with the emergence of Japan as a major manufacturing world power, the general internationalization of competition, a fast rate of technological change, improved communications, a tendency to decentralization, increasing segmentation of consumer markets, and the emergence of the Mergers & Acquisitions specialists. Such factors led many manufacturing companies to restructure, selling off divisions, closing down plants, laying off employees, and continually searching for operating efficiencies.

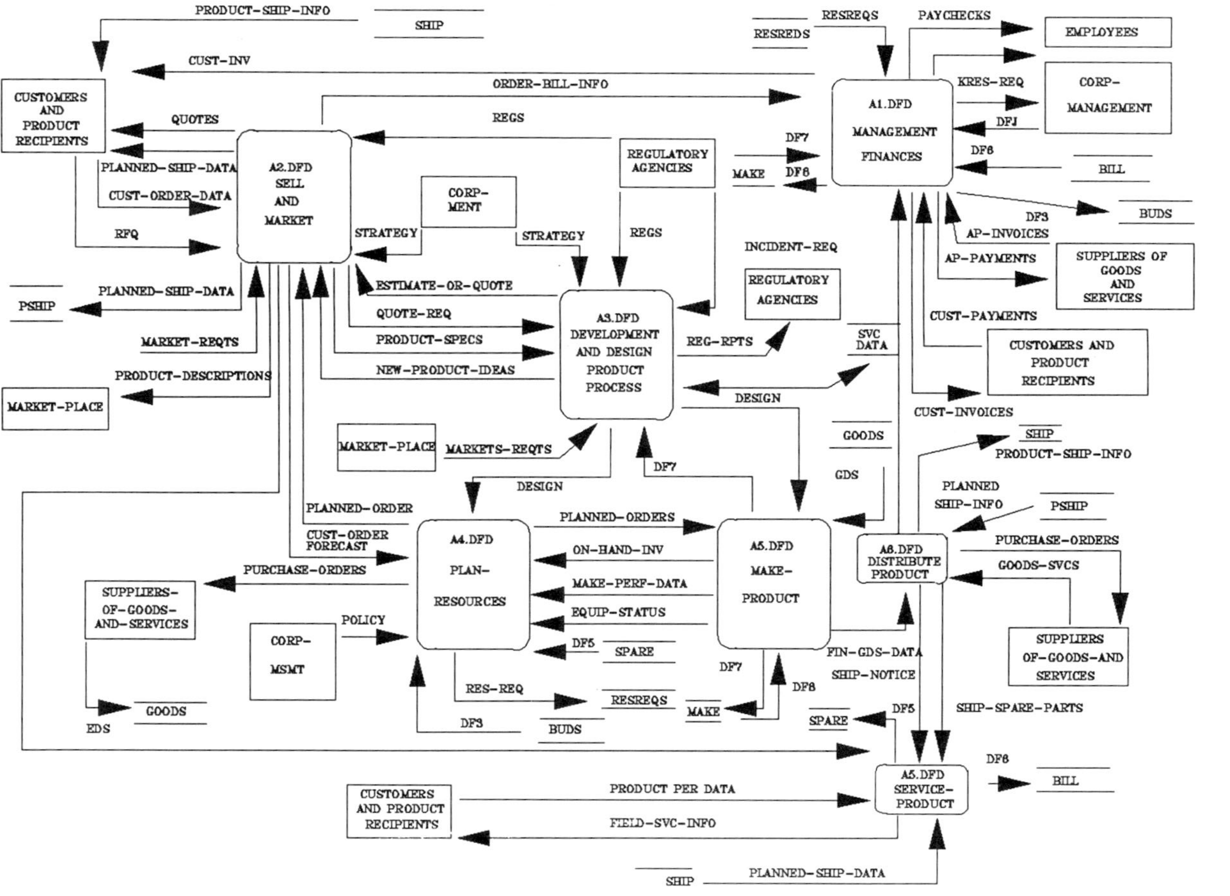

Figure 1.8. Detailed information flow in a manufacturing company.

By the end of the decade, with Japanese manufacturers making massive capital investments and significantly increasing the amount of money going into the research and development of new products, manufacturing companies were under permanent, intense pressure to increase quality, reduce lead times, and reduce cost.

Leading companies offering 5 new versions of a product each year in the mid-1980s were producing 10 new versions of a product each year by the end of the 1980s. Some were able to take 30%–50% out of the cost of a product over a three-year life period. One company reduced order to manufactured product times for basic electronic consumer products from four weeks to two hours. Many electronics companies derived more than half of their revenues from products less than three years old.

The percentage of electronics and computers in all sorts of products was rising rapidly. Customers were increasingly looking for customized products offering more functionality, of higher precision and more reliable, that were at the same time cheaper, resource-saving, and "new." A greater variety of products had to be produced, and the company had to respond quicker to market needs. For the manufacturing company, this implied increasing the quality of today's products and the productivity of today's processes, while simultaneously preparing for more adaptable products and more flexible processes in the future.

By increasing quality, the customer's cost of doing business would be reduced, customer relationships improved, and profits increased. Increasing productivity would reduce the cost of products. The shorter development cycles resulting from increased adaptability would lead to more products getting to market faster. Without increased flexibility, the company would not be able to produce a wider range of products in small batches.

Together, the improvements in quality and productivity lead companies to positions as low-cost producers. The improvements in adaptability and flexibility bring products to market faster, and increase market share. With reduced costs and increased sales, profitability can rise significantly.

In the 1990s, the major issues for manufacturing companies will include the "Information Shock"—the effect of the increasing amount of electronics in products, the possibilities offered by widespread communication networks, and the rapidly decreasing cost of computer power. These imply more frequent design and volume changes, smaller volumes, and much more responsive management.

Competition will increase, not only from traditionally competitive manufacturing companies but also from "corporate raiders" able to show stockholders that they can add more value to the business. As IT-related costs continue to rise, the value of IT to the business will increasingly be questioned by the stockholders.

Faced with this environment, manufacturing companies' prime objective will be to increase their ability to develop new products and services, and find new ways to make and deliver them to the customer faster than competitors. Time, not cost, will become the key parameter. The life cycle of some new electronic products, from conception to obsolescence, is already down to less than two years. As product lifetimes fall further, (e.g. from 24 months to 12 months) the effect of being 3 months late with a product, even if it is cheaper, will be disastrous. Most customers will already have bought the competitor's product. Those who have not will be waiting for the next generation of product. Similarly, producing a product that does not meet customer requirements will be disastrous. There will be no time for trial-and-error, the product will have to be right the first time.

Increasing speed typically requires stripping out unnecessary levels of middle management and bureaucratic control, taking a new look at the whole development to finished goods process, and promoting multifunction teams. Instead of Engineering doing its job alone, then handing over to Manufacturing, which does its job alone and then handing over to Sales, companies will have to bring individuals from Marketing, Design Engineering, Manufacturing Engineering, and Production into a product team with total authority for product functionality, building and costs.

1.7 IT IS ALREADY HAPPENING

The third part of the book gives many examples from various manufacturing sectors of companies obtaining competitive advantage from the use of IT.

There are examples from aerospace, automotive, chemical, electronic, electrical, food and drink, mechanical engineering, pharmaceuticals, publishing, and textiles. In some of the examples, the source of competitive advantage lies within the company. In others, the source lies in the relationships with the outside world of customers, suppliers, distributors, and regulatory agencies.

A few examples will be given in this part of the book of the different ways IT is being used for competitive advantage.

Many automotive suppliers have been pressured into switching from manual order processing systems to computer based methods. Although this has been a difficult and major step for them, it has resulted in inventory reduction and cash flow improvement. It is estimated that further moves to completely replace paper transmission of order, invoice, and notification information by electronic transmission will result in savings of several hundred dollars per car. Suppliers that do not switch to computer-based methods will not be competitive.

The design time of electronic circuits has been reduced, while their complexity has increased. Computer-aided simulation has cut development time by 75%. Using CAD has led one company producing technical instruments to reduce average turnaround time for printed circuit boards from 35 days in the early 1980s to 5 days in the late 1980s. At the same time, the percentage of boards requiring replotting dropped from over 80% to under 5%. Using net-list data from the CAD database helps test engineers to eliminate data entry errors and reduce test development time by as much as 15%. Competitors in this environment who think that the computer's place is in the F&A department do not last long.

Manufacturers of professional electronic equipment have been among the first to equip their sales forces with portable computers running a variety of programs, ranging from current price lists to time management. In some cases, contact time with the customer, a parameter closely linked to sales activity, has risen by 35%.

Traditionally, bakeries have suffered from over-stocking as they tend to cater for worse than average characteristics. From point-of-sale data though, major bakery groups have developed computer models that take into account a range of apparently complex factors, to define delivery requirements. One company feeds weather forecasts into its stock control systems to help identify how customer buying habits are related to short-term expectations of the weather.

Makers of apparently traditional equipment such as turbines and food machinery have equipped their products with computer-based maintenance systems that continuously monitor operating conditions, inform operating personnel of problems and, increasingly, try to foresee major problems that could lead to shutdown.

Development of a standardized computer-based design database for a manufacturer of generator equipment led to savings of up to $100,000 a year per part family. It was found possible to support a reduced number of parts, and to discount aggressively on high volume orders.

A heavy turbine generator manufacturer faced with a declining domestic market and increasing foreign competition found its survival in question. It recognized the use of IT in production as a way to gain competitive advantage. Investing in some 50 CAD/CAM workstations and a similar number of advanced Numerically Controlled machine tools, the company regained its position as a major player.

A toy manufacturer supplied computer terminals to its sales force to help improve customer service and reduce stocks. Once the sales representatives had mastered the system, the company was able to cut its order-entry personnel department by 80%, and reduce corresponding expenses by 70%.

A major player in the global oil-well services industry found its competitive thrust hampered by the lack of a means to effectively manage the flow of design data from engineering to manufacturing and maintenance. It introduced an engineering data management system that, linking 600 users and handling some 300,000 components, cut response time to customers from weeks to hours, yet required 55% fewer engineering support staff.

Some pharmaceutical companies have installed terminals in hospital purchasing departments, linked into their own sales and inventory systems. The hospitals gain by coordinating purchasing and inventory activities. The pharmaceutical supplier gains a privileged relationship with the hospital which soon leads to increased sales.

A manufacturer of personal hygiene products used IT to help handle the problems of distributing a range of over 500 products to several hundred thousand retailers ranging in size from corner stores to big-city supermarkets. Central computers and computers in the warehouses keep tabs on stocks, sales and orders, and helped reduce stocks from 5 to 3 weeks. Trucks are loaded for maximum efficiency. Point-of-sales systems in the larger stores track consumer preferences and are used in predicting demand.

Customers of steel companies can access central computers to make new orders, track order programs, and check their financial position. Electronic data interchange between the steel manufacturer and the customer cuts days from the order placement to the order fulfillment cycle.

In the fashion industry, some middle-market companies can now move clothes off the design board and into shops in less than 2 months instead of 6 months. Orders from shops, with little storage space, are transmitted electronically to computerized warehouse and production facilities. On a daily basis, sales and stocks are examined to define production requirements, and to provide basic marketing information. Bar coded labels help reduce the time to get goods from the factory to the customer by cutting down on inefficient handling and reducing mistakes, such as loading the wrong truck.

Use of the most effective IT techniques allows a greater variety of products to be produced, yet still be profitable over a shorter life time. Textile manufacturers can accept orders for runs as small as 20% of previous minimum yardage, and they can supply customers with fabrics in half the previous time. Textiles, and woolen goods such as pullovers, are designed on graphics screens and data are fed directly to computerized fabric cutting and knitting machines.

1.8 INFORMATION TECHNOLOGY AS IT WILL BE

Information Technology will be used more and more in the future as a way of meeting business objectives, in particular those related to customer requirements. Within companies, the number of IT users will increase. Many will use Personal Computers and workstations connected by local communication networks. Electronic communication with other companies will be more commonplace. Technology costs will continue to fall.

As more internal functions are affected by Information Technology, and IT plays an important role in meeting client requirements, top management will be drawn more and more into IT issues. Some top managers will try to avoid these issues as long as possible. Others will understand that by taking action early, they will gain competitive advantage. They will realize that this cannot be achieved only by using IT technology made available by vendors, or by using IT in the way that is most interesting to the EDP Manager. Managers from all departments will have to be trained to make better use of the possibilities offered by IT.

Companies will have to understand where and how they can beat the competition and focus their resources on these areas. Where IT can be used for competitive advantage, companies will be willing to invest. Where the use of internal IT resources is not strategic, there will be an increasing trend to use external resources.

The major change for top management will be to become involved in setting the IT strategy, in particular, as a function of the overall business strategy. The way a given company will use IT will depend on the company's particular circumstances, but some overall directions can be foreseen.

Looking outside the company (customers, suppliers, regulating bodies) the first issue will be communication of information. This will take different forms (a terminal outside an organization accessing a computer in the organization, information being transferred from a computer in one organization to a computer in another organization). The information will be in a variety of forms—voice, text, graphics, numeric.

Looking inside the company, the communications issue will also be important. Product teams, pulling in members of different departments, often in different locations, sometimes in different companies will increasingly communicate electronically rather than on paper. Similarly, sales representatives in the field will want to be in close contact with information in the company. Software that is closely tailored to individual user's needs will be required for all departments—from marketing and sales, through engineering and manufacturing, to more administrative roles.

Underlying the communications and information processing requirements will be a major requirement for software that can effectively and efficiently manage the vast volumes of information that will be both needed and available.

Given this apparently purely technological background of information communications, processing and management, one may wonder how a company can aim to win competitive advantage because all these technologies will be widely available. First, many companies will not believe that IT can help them to attain a competitive advantage, or they will believe it but not be able to implement it. Secondly, even in areas where it is possible to copy there will be a major advantage for the first company that succeeds in using IT for competitive advantage in a given role. The innovators are more likely to be the winners than the followers.

Third and perhaps most important, is that to attain real, long-lasting competitive advantage technology alone is not sufficient. Organizational issues are just as important, and whereas technology can be bought, organizational changes take place very, very slowly. By adopting a competitive advantage approach that relies on both technology and organization, a company can gain competitive advantage. This implies of course, that such advantage cannot be gained overnight. It has to be planned for in the long-term and implemented over a period of several years. The plan is no longer an "IT plan" but an "IT-related business plan." A five-year plan of this type can lead to a three-year competitive advantage, not because competitors cannot buy the technology but because their organizations are not able to use it effectively.

People will look in various areas to gain advantage from a technology. In the case of railroads, for example, many more were involved than just those manufacturing and laying the rails. Locomotives, passenger carriages, and specialized goods wagons had to be built and equipped. Property speculators became involved as the coming of the railroad opened up new towns. The oil industry expanded as it could now move products in bulk. Cattle were transported from distant southern ranches to the plates of the huddled northern masses. In the same way, competitive advantage will be gained not only by those who manufacture the basic IT components, such as computers and communications equipment, but also by those who see the opportunities that IT will create for society as a whole.

2

FROM BUSINESS STRATEGY TO IT STRATEGY

"Nobody wants to hear Drucker's Law that management is 99% conscientiousness and only 1% genius, and you can do without the genius if you're really conscientious. It's hard work, conscientious work. You don't cut corners, and you don't grandstand, and you keep it simple. And work and work and work. But once you accept this, we can now use information in a defined, logical form that can be manipulated and presented, displayed, and analysed. For strategic issues within a system, we can use information as a help in making decisions about the system, but we cannot yet formulate the decision in information terms about the system itself. In part because we haven't done enough work. We do not have the foundation.

All this changes the definition, and that is what information systems cannot handle. In old Aristotelian logic there is a lovely distinction between imminent and transcendent problems. Problems within what we call a system and problems from without. The transcendent problem changes the meaning of the term within the imminent. So when you have social and economic events, the most important event is irrelevant. The unique event is a freak, an exception, and in social and economic systems, the unique event changes the system. It's a Henry Ford, having the great benefit of not having gone to college so that he never had to unlearn economics. He didn't know that you make the most money by producing the fewest with the highest price. And that changes the system, because suddenly it changes the meaning. Certain people see that and others don't, and the ones who see it cannot teach it. I'm good at it, I'm born with it. But I've never been able to teach it."

Peter F. Drucker in *Handbook of MIS Management,* Second Edition, Robert E. Umbaugh, ed. Auerbach, New York, 1988.

2.1 ONLY TOP MANAGEMENT CAN MAKE IT HAPPEN

Top management alone can link IT to the business mission and can make sure that due account is taken of IT when the business strategy is defined. They alone can ensure that the necessary changes occur throughout the company. Sometimes it may be that someone else in the company or an individual department will develop an IT activity that offers competitive advantage, but even in this case, for real benefits to occur, top management must support the development, make sure that it is focused properly, and that developments in other areas of the company are brought into line.

In the past, in the MIS—EDP environment, top management did not take the lead, preferring to delegate it to the EDP Manager or to individual functional managers: the results are clear to see. IT appears to be a costly and low-quality support function, not a technology that can and should be used to improve the standing of the company in the market-place and improve customer perception of its products. That this state of affairs has occurred is the fault of top management, and it is now up to top management to put things right. The people who have been running IT have not been looking outside the company to understand customer's needs. Instead they have been looking inward at the internal parts of their computer systems and "watching" the computer manufacturers. To a certain extent, this should be done but it is a lower priority activity than matching IT resources to client requirements. Perhaps it is not entirely their fault, because top management may not have invited them to participate in the most strategic activities.

It will now be a task for top management to get the company to use IT for competitive advantage. One percent of the process will be in finding the inspiration, the flash of genius, that it can be done, and how it can be done. The other 99% will be the hard work of developing the strategy and then managing the planning and implementation phases. Although the major aim of the book is to help top management find the 1%, it will also help with the other 99%—particularly in communicating the message that IT is to be used for competitive advantage and in developing the corresponding culture. Once this book has succeeded in convincing top management, it can help spread the message throughout the company.

IT can affect all the company's activities and beliefs—even the business mission. In most companies, the business mission will already have been defined, but it may need to be revised as a function of the potential of IT to meet client requirements in the future environment. Computers and communications can change the business, affecting products, services, and processes. The business mission and objectives need to be reviewed at the highest level of the company—by the Board of Directors and the CEO. The Board and the CEO need to be aware of what IT can do for the company. They will need to foresee the new environment that the company is moving towards, agree on this picture, and start to prepare the company for it.

Without such an understanding, the chosen business mission cannot be defended. A suitable definition of the business mission and a complete understanding of the effect of IT on the business objectives are critical to the processes that follow later. If they are not attained, the result may be that the wrong products will be produced, the wrong markets addressed, and the wrong organizational structures developed. This responsibility will lie with top management.

In a relatively mature business, it might seem unnecessary for the Board to examine the business mission as a function of IT. After all, for the last twenty years the company has probably been making the same type of product with fairly similar processes for the same customers. Is it reasonable to claim that all this might change, just because someone thinks up a new use for a computer? In the past it has happened. Companies in mature business sectors have disappeared because they were not prepared to accept that change might occur. Most mature companies, it is true, will not need to fundamentally change their missions. However, what about young companies or start-ups? For them, almost the opposite is true. They can hardly start their existence in the business world without taking account of the effect of IT.

Who in the company will be defining the business strategy? The top management team should be, unless they feel that this is something that can be delegated. Even if it is delegated, the team still has to make the final decision, and in today's competitive world with takeover specialists looking over their shoulders, they will not want to take the wrong decision. To take the right decision, top management must understand what IT can do for the company and what effect it could have on the company's markets, products, and processes. Based on this understanding, they can gain a vision how the company can benefit from the use of IT in the future.

Is it really the role of the top management team to get involved with computers and the future of IT? Surely top management should be involved in more important issues, such as modifying last quarter's budgets, tracing mislaid stocks, or checking up on the number of personal telephone calls made by sales staff? Shouldn't top managers leave the strategic IT questions to the specialists? After all, there are an awful lot of specialists on the market, all claiming they can help top management use IT to gain competitive advantage. There are strategy consultants, management consultants, technical consultants, IT consultants, IT strategists, system integrators, application systems developers and vendors, computer manufacturers and vendors, manufacturing automation consultants, engineering contractors, software houses, integrated systems vendors, automated manufacturing systems vendors, project management specialists, and training organizations.

With so many specialists to choose from, is it really necessary for top management to do anything but pick the right specialists and then sit back and wait for success to happen? The real problem is that though the specialists might know "how to do it" they don't know "what to do" at the top management level, they cannot lead the rest of the company, and they cannot make high-level decisions. They cannot lead the company towards competitive use of IT any more than the company's own EDP analysts, order processors, or shop-floor workers. The only possible leader is the company's top management team. These are the people who know the company best, who understand the customers, who know how the employees will behave, and who have a detailed knowledge of the products. They are the only ones who can take the high-level decisions, make sure that objectives are met, that funding is found, and that the organization is modified where necessary. Certainly, they will want to make use of outside assistance at various times—possibly with help in defining the business mission or in writing computer programs. They cannot go outside and expect to get leadership and detailed knowledge of their own company. It is their responsibility to start the chain of activities that will lead the company in gaining competitive advantage through use of IT.

Long-term competitive advantage through the use of IT will only be sustained by the right blend of organizational and technological factors over a considerable period of time. Top management cannot just agree to a strategy put forward by a team of external specialists and hope that the strategy will work itself into reality. They must communicate the strategy, making sure that realistic plans are developed, and that these are implemented over, in most cases, a period of several years.

Provided that the strategy takes a flexible approach to technology, the major problems in gaining and keeping competitive advantage will be organizational. At the heart of the problem lies the everyday behavior of the various people within the company. Many middle managers will want to preserve their functional empires, others will clearly not want to lose their status. Many will try to prevent the free flow of the information which in the past they have jealously guarded and kept from the very people whose business success depended on it. The traditional EDP specialists will resent what they may see as their diminished role. As this type of problem can be foreseen, top management can take prior action to prevent it.

In the following sections of this chapter, it is discussed how business planners have ignored the IT opportunity, and EDP planners have ignored the business environment. It will then be shown how top management must overcome these problems, take command, and lead the company towards developing a strategy and organization that will support the use of IT to gain competitive advantage.

2.2 HOW TRADITIONAL BUSINESS PLANNING IGNORES THE IT OPPORTUNITY

Once the members of the top management team of a manufacturing company have understood that it is possible, and perhaps necessary, to use IT to gain competitive advantage, their next step will be to try to put theory into practice. They will want to know how to do this and will look for methods that other companies have used successfully. They will soon find that there is no standard methodology that works for all companies and that the best approach is to combine parts of available methodologies that are most relevant to the company's particular circumstances.

The methodologies that are available come from two sources—traditional business planning and traditional EDP Planning. The majority of business planning methodologies currently in use were developed prior to the 1970s, well before the potential strategic value of IT was fully recognized. As a result, they do not specifically take account of IT. Most of the EDP planning methodologies currently in use address the development of software and fairly small-scale systems. Some address larger-scale systems and IT strategy but generally they do this from an EDP or functional efficiency point of view. They do not relate the use of IT to major business objectives in the competitive environment.

The business planning and EDP planning methodologies should be brought together so that IT strategies can be developed as a function of business goals and their implementation will lead to the goals being met.

This section of the book outlines the classical business planning process and important additions to it: then the classical EDP strategy, system and software methodologies are introduced. The descriptions of the business planning and EDP methodologies lead into an introduction to an approach for developing an IT-related business strategy. Finally, the organizational impact of using IT for competitive advantage is addressed.

Top managers will probably find nothing new in the section on business planning and EDP gurus will find nothing new in the section on EDP planning. Each group of readers will therefore find one section superfluous—although to meet the aims of addressing top management and then helping to communicate the message throughout the company, both are necessary.

The term 'business mission' is used to describe the highest-level purpose of the company's existence. The business mission is a statement in, at most a few hundred words, of the purpose of the business. It may outline products and services, markets, channels through which markets are reached, means of financing and expected financial return, and any organizational and policy issues that may be particular to the business. Such a statement should be meaningful over a long period of time—ten years or more. It serves as a foundation on which shorter-term decisions can be taken. The highest-level targets that are set for the company are referred to as the business goals. The business mission and business goals express, at the highest level, what the company wants to do. Responsibility for their definition lies with the owners, or the representatives of the owners, of the company.

The business mission has to be converted into a set of operationally effective business objectives relating to the major functions of the company. There are three major business objectives, addressing the key issues of marketing, innovation and integration. The marketing objective looks outward into the market—identifying customer needs that the company should meet. This leads into a definition of who the customer is and will be. The innovation objective is more inward-looking. It is product-oriented, aiming to identify products that the company can produce, and that will be superior to competitive offerings. The integration objective keeps the marketing and innovation objectives in balance. It addresses the company's resources and aims to ensure that real market needs are met by a competitive product that can be produced in competitive time and cost.

The business strategy will express how the company will meet these objectives. The strategy will show how the company's resources are to be organized and make clear the policies governing management of the resources. Responsibility for the definition of the strategy lies with the top management team. They may delegate part of this task to others and will delegate the implementation of the strategy to others, but they must, nevertheless, be involved in the process, providing important inputs and controlling implementation. The strategy should be set for a 3–5 year period.

The business strategy should be reviewed annually during its lifetime and may need to be modified. However, it should only be changed for very good reasons. Changing the strategy after just one year implies that it was not set properly in the first place—and this is a very expensive affair.

Once the strategy has been defined and communicated, the detailed planning can start—again looking at a 3–5 year period. This will be carried out by middle management in line with company policy. The best way to implement the strategy will be decided. The following questions should be asked:

- How will each function be organized?
- What activities will have the highest priority?
- Who will manage each activity?
- How will the activities be financed?

When the detailed plans have been developed, the implementation process can start. Because this process lies outside the scope of this book, it should not be taken to imply that it is not important: it is of the highest importance, and many books have been devoted to the subject. In practice, top management will often be able to identify a suitable strategy quite easily, (part of the 1% of genius) but will find its implementation to be much more difficult (part of the 99% of conscientious work).

The classical business strategy definition process takes the business mission and goals as a starting point. It is a four-step process, starting with the preparation of basic information, and following through the steps of determination of the company's resources, formulation of potential strategies, and selection of the chosen strategy (Figure 2.1). In the first two steps, the intention is to gather and agree on the basic information that will form the basis for developing potential strategies. Information on markets, products, and company resources will be collected. The following are important business strategy development questions that must be answered:

What are the major market segments?
What are their niches?

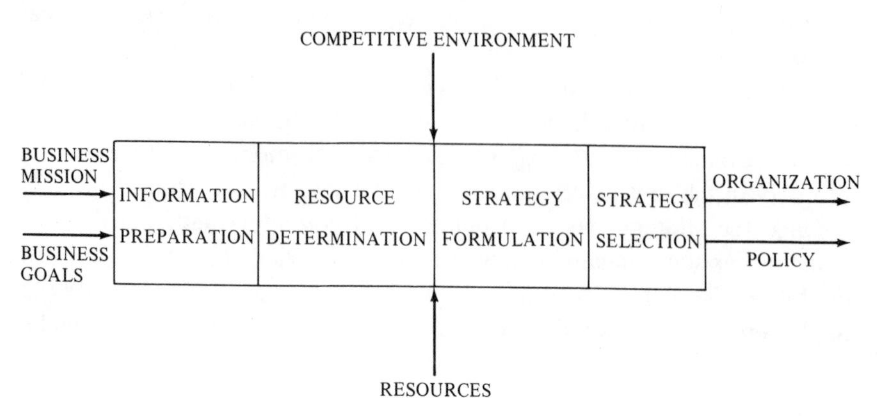

Figure 2.1. The Strategy Development Process.

What market share will the company need to be dominant?
Who are currently the company's clients?
What are their needs?
Who will be the company's clients in the future?
What will their needs be?
Will the needs change?
If so, how and why?
Where will the company's customers be?
How will they buy?
What will customers see as value?
What will be key purchasing criteria?
Who are the competitors today?
What are their plans for the future?
Who could be the company's competitors in the future?
Where is the company strongest?
Where are the company's competitors weakest?
Which new markets could the company enter?
What products does the company currently make?
What, if anything is unique about them?
Where are they in their life cycles?
Are they embryonic, growing, mature, or aging?
What will the company produce in the future?
What will be unique about these products?

In what volumes will they be sold?
What will the company make itself, which parts will it buy in?
What are the sources of supply?
Should the company collaborate with other companies on R&D?
What are the current resources of the company?
How are they organized?
Why are they organized like this?
What technology is being used?
Is it only used to replace people or also to add to their value?

Once the basic information has been assembled and agreed upon, there will be several possible strategies. Each must be examined in detail to show how it would lead the company towards meeting its business mission and goals. For each possible strategy, the links to the strengths and weaknesses of the company and the opportunities and threats in the market, need to be identified. Outline functional strategies should be developed to show how activities would be carried out. The way in which the organization would be structured and the relationships between the various parts of the structure should be reviewed.

During this process, which can be quite lengthy, it should become clear which are the most suitable strategies under review. There may be one or more strategies that meet all the requirements, in which case the best should be selected. The results of the strategy definition process will show how the company's resources are to be organized and managed. The strategy will define which products and services are to be offered and the markets to which they will be addressed. Forecasts will be made for revenues, income, expenditure, and cash-flow. Investments and capital requirements will be estimated, and functional goals and strategies outlined.

The classical business strategy definition process outlined above offers a basic framework for developing a strategy. It does not lead a company to take a particular stance; for example, it does not lead a company to use IT to gain a competitive advantage, neither does it push a company to try to be a dominant, long-term market leader, show a company how to use IT, nor identify measures to indicate whether IT is being used effectively. It does not address the question of successfully implementing the chosen strategy nor the question of getting the entire work-force to work whole-heartedly towards meeting the business goals. To overcome some of these deficiencies, the following additional methodologies have been proposed.

There is a set of methodologies based around identification of 'Success Factors' and implementation of corresponding plans. A Success Factor is defined as something that must occur if the business goals are to be attained. The methodologies are based on practical experience demonstrating that it is effective to get top management to focus on a limited number of critical issues that must be successfully managed if the business is to succeed. The approach starts with the goals of the business (e.g., to be a player in the five major global markets of North America, Western Europe, Japan, Pacific Rim, and Latin America; to be in the top three by market share in each market; to have a particular Return on Investment). The Success Factors necessary to meet these goals can then be grouped under the relevant heading—Quality, Productivity, Adaptability, or Flexibility. The factors relating to current products will be grouped under Quality, and those related to current processes under Productivity. The factors relating to future products will be grouped under Adaptability, and those related to the processes associated with future products under Flexibility.

Identification of the Success Factors will require the participation of many people and will be a lengthy process. Each person will initially identify a particular, personal set of Factors. The Factors identified by people in different parts of the company will then need to be brought together (Figure 2.2) and arranged in order of importance. The relationships between them must be clarified. An agreed list containing the most important Success Factors (probably between eight and twelve) can then be drawn up. This list will indicate some of the areas where IT may help support the drive for competitive advantage. It will be useful in setting up a direct link, and closing the loop, between business goals, Success Factors, Measures of Success, business strategy, IT strategy, areas in which IT can be used to gain competitive advantage, and the selection of the corresponding systems (Figure 2.3). For a car manufacturer, the factors grouped under the heading of Quality may point, among other things, to the use of CADCAM. Those grouped under Adaptability may point to Telecommunications as a way to help Marketing better understand customer needs.

Another set of methodologies is based around value chain analysis—a technique which recognizes that a company is made up of a set of discrete activities, such as logistics, sales, production and service (Figure 2.4), each of which is part of the overall chain of activities that add value to a product. The company's value chain can be considered a system of interdependent activities, each of which contributes to the company's overall position and

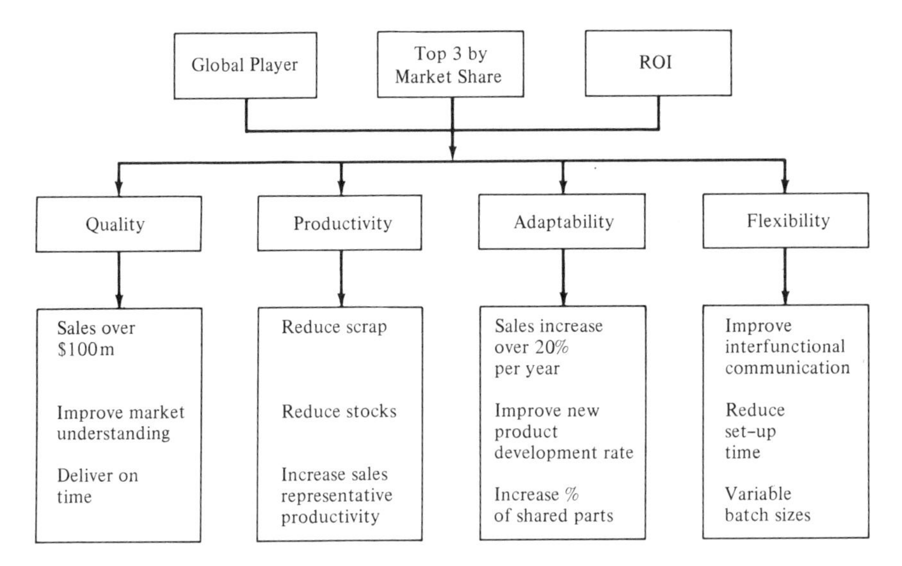

Figure 2.2. Business goals and Success Factors.

offers potential for differentiating its products from those of the competition. The company's value chain extends outside the company, where it meets those of its suppliers and customers (Figure 2.5). By including competitors in the picture, linkages between activities and other sources of competitive advantage become apparent (Figure 2.6).

From the IT point of view, the overall value chain into which the company fits should be analyzed both as a whole and as three interlocking entities. The three interlocking parts are the part on the supplier side of the company, the part that represents the company itself, and the part on the customer side of the company. Analysis of the three parts could show opportunities for competitive advantage in, for example, incoming logistics, engineering, and sales, respectively. However, such an analysis is not sufficient as it would ignore the opportunities for the company to use IT to help reposition itself in the value chain; for example, by taking over some of the functions previously carried out by a distributor, or even by the final customer. As a result, analysis of each of the three parts should be complemented by an analysis that examines the overall value chain.

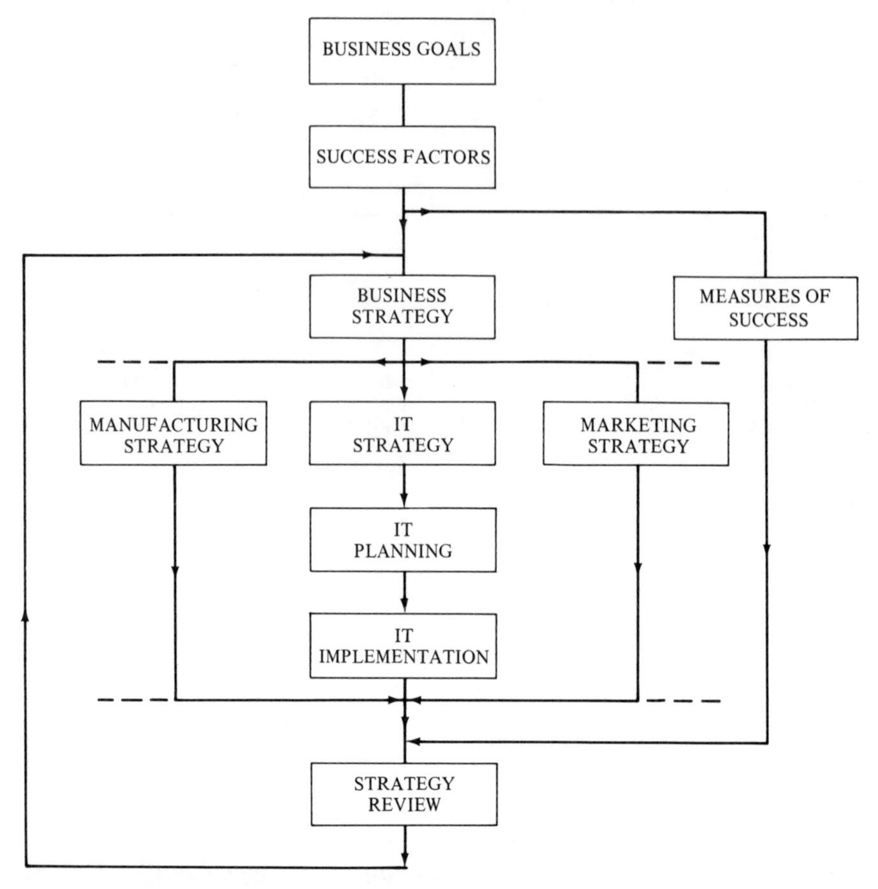

Figure 2.3. Closing the loop between strategy and implementation.

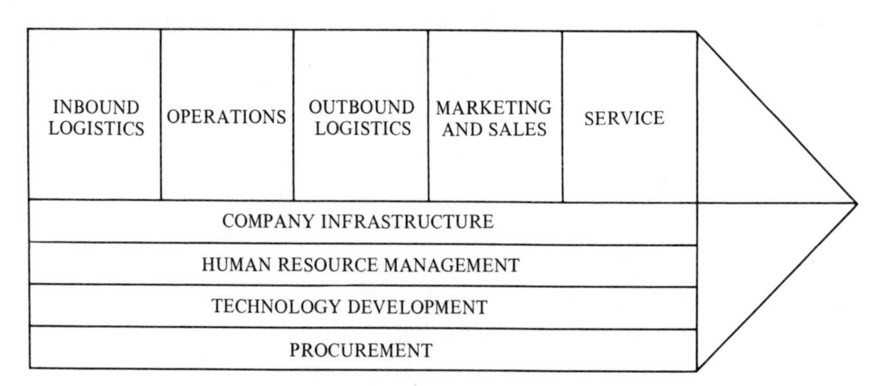

Figure 2.4. The Value Chain.

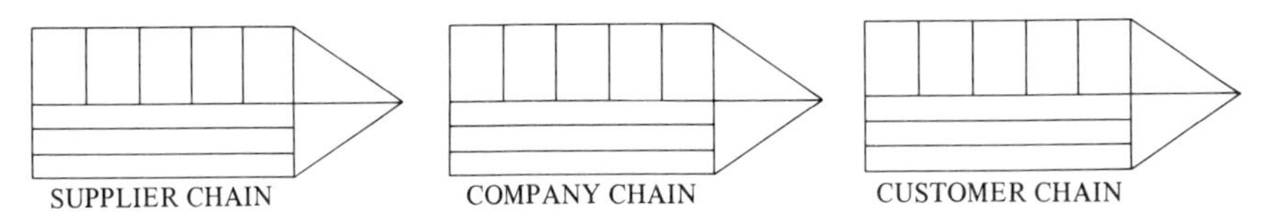

Figure 2.5. Linking Value Chains.

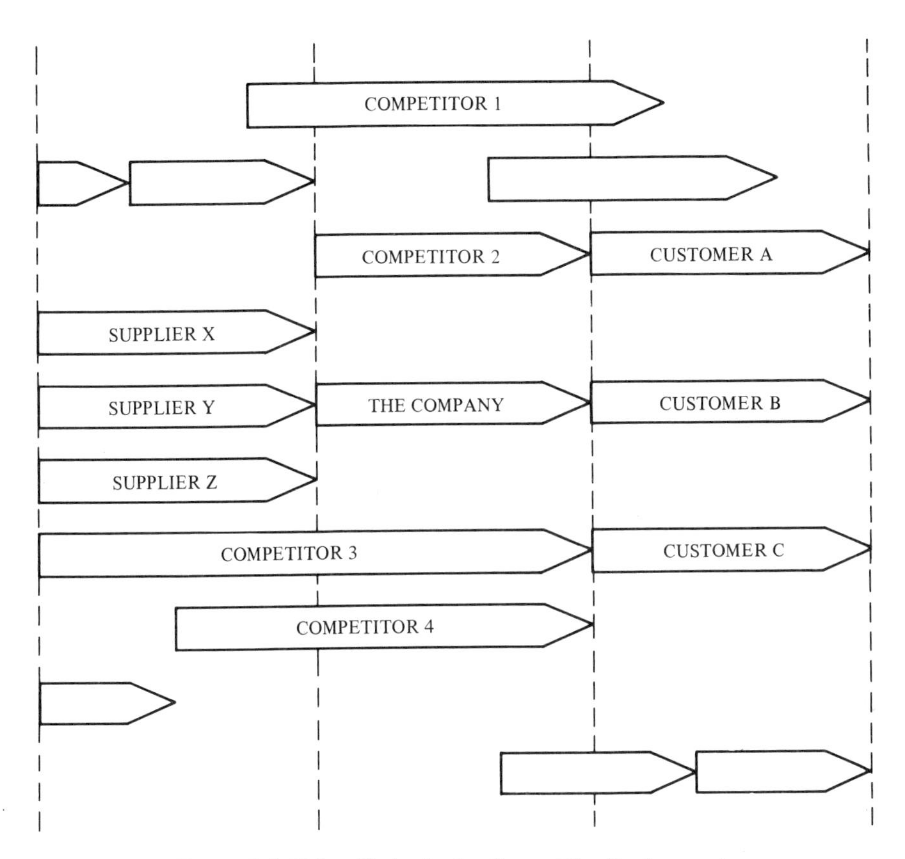

Figure 2.6. Value Chains in the Competitive Environment.

2.3 TRADITIONAL MIS—EDP STRATEGY DEFINITION METHODOLOGIES

In the past, just as business planning methodologies did not take account of IT issues, the EDP strategy definition methodologies did not take account of the business issues. As a result, it is perhaps an exaggeration to refer to them as EDP strategy definition methodologies, since in many cases they are closer to being EDP planning techniques. They were developed to meet the need created by the availability of EDP and the lack of top management guidance as to its most appropriate use.

Various methodologies have been developed to assist in the definition of EDP strategies and the implementation of corresponding systems. Some cover the whole range of IT activities, others are limited to just one part of it. Most of them can be classed into one of three categories. The first category, the EDP strategy definition methodologies are discussed in this section. The other categories, system development methodologies and software development methodologies are discussed in Section 2.3. There can be considerable overlap between the various phases of the three methodologies.

EDP strategy definition methodologies have developed over the years in a similar way to the development of the use of EDP and MIS. There are many methodologies relating to the use of EDP and MIS in the traditional MIS—EDP environment, but most of them ignore the competitive issues. They tend to be very inward-looking, concentrating on the company and the EDP activity itself, and ignoring the external world of customers, suppliers and competitors. Only recently has the need been recognized for methodologies that address the use of IT for competitive advantage.

In Section 1.3, it was discussed how MIS—EDP had evolved, how their use had evolved, and the way in which the focus of activities has changed. It was shown that there have been the following phases:

- automating manual activities
- improving business efficiency
- using MIS—EDP efficiently
- spreading MIS—EDP throughout the company
- integrating MIS—EDP activities
- using IT for competitive advantage

Against this background of changing phases, individual companies have followed their own IT development path. Some have been very advanced, trying to use MIS, EDP, or IT in what appears to be the most suitable fashion at a given time. Others have lagged behind, perhaps because they did not have the means to invest, they did not agree with the market trends, or they did not know what to do. The way that companies, at the two extremes of the spectrum, will approach IT now will be very different. The advanced IT company will have some sort of an IT culture, even though it may not be the right culture, and many of the associated issues that will be raised by investigating the competitive use of IT will have already have been identified, if not resolved. On the contrary, in the company that has not been making major efforts to make the best use of IT, the personnel and the organization may just not be able and ready to support use of IT for competitive advantage. As a result, before a company tries to use IT for competitive advantage, it should understand its current position and capabilities and its overall intentions. Only then can it try to pick a suitable IT strategy definition methodology. The methodologies are different, assume different starting points, and tend to lead to certain fairly-well defined results. Because, in a way, the choice of methodology defines the type of result that will be generated, it is important to pick the right type of methodology. The wrong type of methodology may never lead to the right result for a particular company.

One of the simplest types of methodology is based on the hypothesis that companies follow similar phases of MIS—EDP growth. Under such a hypothesis, a company that is not a leader in the use of IT would find it useful to compare itself with a more advanced company. It could identify the phases that it has not yet experienced, and plan to carry out the corresponding actions as efficiently as possible. Such a methodology will be of little help to advanced companies. Neither is it likely to show where a company can gain competitive advantage. A third weakness of this methodology is that it assumes that all companies, regardless of sector and strategy, will develop IT along the same lines. As companies increasingly look to develop differentiated niche strategies, this will become even less true.

Another often used methodology involves tracking the use and flow of data in the company: this can lead to a detailed understanding how that data is used in the various departments of the company, and the way that data flows between departments. At the heart of the methodology lies the belief that if the company understands very well what it is currently doing, it can probably find ways of doing it better in the future.

In a company of reasonable size, the effort required to identify and examine all the data flows can be enormous. Many companies that have tried to carry out such an exercise have had to limit the degree of detail of data that is collected, otherwise they would never have finished the task. Limiting the scope of data collection in this way can be dangerous. By not investigating data in other organizations, such as suppliers, who may share data with the company, or by not looking at the use by other organizations of company data, the value of the results—particularly from the competitive point of view can also be reduced.

Companies that carry out the exercise in detail generally find they learn more about how their company works. They see where data flow is disorganized and data is duplicated. The exercise can provide a good basis for the definition of the best flow of information through the company. In turn, this can be a solid framework on which applications can be built. The question arises though as to what the 'best' flow refers to. Is it the best way for being a leader or a follower? Because it is the current use and flow of data that has been examined and resulted from a business strategy which did not take account of IT, the "best" way may not be relevant to a new strategy that incorporates IT. The whole activity may be superfluous, with the net result after two years of study being a 3-inch thick "IT Strategy" that is immediately obsolete and will be quietly ignored a few weeks later. This misguided use of resources results from an initial failure to identify the correct objectives for the study.

Rather than taking this type of data analysis approach, with its limited objective of increased efficiency, it is more useful for companies to take an almost opposite approach, and develop an IT-related business strategy taking account of competitive issues. From this, they can understand what data is required, how it should be organized, and how it should flow. Because IT is so unlike EDP and MIS, the resulting data requirements will probably be very different.

A third type of strategy development methodology is based on an application portfolio approach. Such an approach maps out all possible areas of the company where applications could be used, and for each area investigates whether an application is in use, and if so, what it is being used for, and who is using it. This approach gives a good idea as to what the company is actually doing, and what it could be doing. This "application-tracking" approach, like the data-tracking methodology previously mentioned, tends to be reactive and is not likely either to show that a particular company is not addressing IT in the right way, or to show which would be the best way to use it. An extension to the methodology to include a comparison of applications used by the company with those used by competitors can be helpful in identifying potential weak spots but again will not show how the company should use IT. The approach to IT needs to be proactive not reactive.

A fourth type of methodology focuses on the "integration" of the various applications that are already in place. Integration will become an issue when there is already a large number of basic applications in place. In its simplest form, integration only aims to allow smoother communication of information between applications and to improve the efficiency of the systems in use. In this form, it does not specifically direct the company towards using IT for competitive advantage. Rather than identifying areas requiring peak performance to support attempts to gain competitive advantage, it is more likely to lead to lowering the level of high-performing areas and raising the level in areas where there are performance problems. Once this danger has been recognized, it can be avoided. Unless one is careful, integration, by its nature, will be reactive and build on the past, rather than being proactive and look to what may be a very different future.

A fifth type of methodology, "application portfolio management" aims at obtaining the maximum possible benefit from the available MIS—EDP investment and the applications currently in-place. Once the applications that are needed to help gain competitive advantage have been identified, the methodology can be of use in helping to focus resources on them. Application portfolio management can help to build up a well focused set of applications, because adding new applications that are closely related to existing applications often has more effect than developing new applications in areas where there are currently none.

Whichever methodology or mix of methodologies is used, the aim should be the same—to understand and define how the company will use IT. The result should be made available to top management in the form of a concise report. This should explain why a certain course needs to be taken, what this involves and, at a very superficial level how it will be implemented. It should contain an overview of the architecture—the overall layout of all the IT (computing and communication) resources. It should discuss what hardware will be used, where it will be used, and how the components will be connected. Similar pictures should be available for the software, the data, and the human resources. The direction chosen for the use of IT should be explained and the major policies and priorities governing management and use of resources defined. Also the role of IT within each function should be addressed, and the expected results described. The organization of IT resources should be defined, for example, showing the structure of the IT function itself, and the way it will work with the other functions. Any other important issues, such as the use of international standards, the policies for recruitment and training, and the methodologies to be used, need to be addressed.

In practice, the report that attempts to define an EDP or MIS strategy rarely meets this aim. Instead it is generally very long, and rather than focusing on what needs to be done and why, mainly addresses the issue of how to implement a piece-wise continuation of traditional MIS—EDP behavior. Detailed project plans and data flow diagrams abound in the report along with much discussion as to why one computer was chosen in preference to another. Long lists of available application packages are also included.

The link between the overall business strategy and how IT is to be used, will probably be missing from such a report. In the absence of this information, any measures related to its use will not be expressed in terms of meeting business goals but in terms of internal characteristics, such as a percentage increase in budget, the number of terminals in use, the computing power available, or the number of lines of program produced each day. Generally it is only when a methodology aimed at helping the company gain a competitive advantage through IT is used that measures related to achieving business goals are developed.

2.4 SYSTEM AND SOFTWARE DEVELOPMENT METHODOLOGIES

Many methodologies have been proposed to help in the expensive, complex and time-consuming task of systems and software development. Whereas it is useful for top management to understand the strategy development methodologies described in the previous section in some detail, the same level of understanding of system and software development processes is not needed. However, it is important for the top management team to be aware of some of the methodologies in use so that they will have some understanding of the discussions and reports concerning the implementation of their IT strategy. Much of the detail will appear unnecessarily technical and low-level to top management—and it is. Top management should set the strategy and make sure that the desired results occur. They should not be involved in the details of planning and implementation, which is where the system and software development methodologies belong.

If most communication between the IT function and top management takes place at this level, then something is wrong. Top management should communicate at the strategy level with the leader of the IT function. This person will communicate at the system development and software development levels with the IT specialists, and, when appropriate, with users of IT.

There are two major classes of system development methodology.

One takes a life-cycle approach, the other involves prototyping. The system development life-cycle approaches assume that the life-cycle can be broken up into a certain number (often 7) of discrete, sequential steps. These include the following:

- Feasibility
- Preliminary Design
- Detailed Design
- Programming
- Testing
- Implementation
- Maintenance

Depending on the approach, the steps may well be given other names, such as:

Approach 2:

- Project planning
- Requirements analysis
- System and procedure design and development
- Product evaluation and acquisition
- System and acceptance testing
- Transition
- Maintenance

Approach 3:

- Start-up
- Feasibility analysis
- Requirements analysis

- System design
- System development, conversion, and integration
- Installation and testing
- Operation and maintenance

Approach 4:

- Project definition
- Functional specification
- Functional design
- Detailed design
- Implementation and integration
- Field start up and commissioning
- Maintenance

Although the names may differ between the approaches, the activities covered are very similar. Most of these approaches were developed before IT was used to gain competitive advantage, so it is not surprising that the first steps may overlap with part of the IT strategy definition process. At that time, rather than implementing an IT strategy, the EDP Manager would have been looking for systems to develop, and the "project definition" step was needed to help understand what was being proposed.

Similarly the "feasibility study" step may be less important when the system to be developed has been agreed upon by top management to be necessary to gain competitive advantage. In the past, it was necessary to show that a proposed project offered some benefit. This step is also helpful to choose between competing tactical systems. The feasibility study generally involves a review of existing systems, an overview of requirements, a review of potential solutions, and a preliminary evaluation of the financial, technical, and organizational feasibility of the recommended solution. Outline planning, structure, constraints, objectives, orientation, and field of study will be defined for the next step of the life-cycle. In some cases, a feasibility study may show that some specific short-term activities should be taken before all the steps of the life-cycle have been addressed.

The requirements analysis determines the detailed business needs that the proposed system should meet. It defines the detailed functional requirements of the users. A high-level model is developed describing how the current system works. Special interface, processing, and other characteristics will be identified and evaluated and the best solution will be recommended.

At the system design step, a detailed specification of the system will be prepared and a general systems framework developed. This will show how the system is split into major subsystems. Modeling will be extended to lower levels, and function charts, information flow diagrams, and other documents developed to specify the subsystems and their constraints. The requirements for modification of existing systems, development of new systems, and the integration of the two are defined. More detailed financial analysis is carried out. User interfaces, such as report layouts and screen presentations are defined. Procedures for using the system, procedures addressing error-handling, back-up, and security are specified. Now the system and the system architecture are defined in detail. Hardware, software, and communications components are precisely specified. Processing logic, data structures, input/output requirements, system and user procedures will be detailed. Suppliers are chosen, and equipment and sub-systems ordered. Detailed system delivery and acceptance tests will be specified.

Programs are developed and tested, first individually, and then together. Other components of the system are commissioned, and the various parts integrated and tested. Existing systems are modified where necessary. Users and operators are trained to work with the new system, and technical documentation is consolidated.

The system is installed, brought on-line and tested, then system and user procedures are tested. Test data is loaded and the whole system run in the everyday user environment. Finally, the system is formally accepted: then the maintenance process begins. This involves correcting any errors and making required improvements and is an activity that will continue throughout the rest of the life of the system.

The system development life-cycle previously described is logical and successful in environments where it is possible during the early steps to determine user requirements and where these do not frequently change. However, this is rarely the case. Another problem that arises when the life-cycle approach is taken is that users are often unable to communicate their real requirements—or at least the system developers are often unable to understand them. User requirements often only become clear once the system is already in use, and users have a better understanding of what it could do. As a result, rather than following a straightforward linear path through these steps, the system development life-cycle often involves two steps forward and one or more backwards. The maintenance activity, rather than being simple and keeping a system that is 99% in good working order, becomes the major activity of the life-cycle, involving redefinition of requirements, redesign of the system, and redevelopment of system components and procedures.

To overcome these problems with the life cycle approach for system development, the prototyping methodology was developed. In this iterative approach, it is accepted that user requirements cannot be fixed at an early stage and will change significantly as users get to know a system. As a result, instead of going through the "life-cycle" and waiting many months or years for results, the user can request that a prototype system be built to meet requirements. The user then evaluates the prototype system, which is then revised to take account of the redefined requirements. The process is repeated until a suitable prototype has been developed (Figure 2.7). Prototypes can be developed to different levels. Sometimes a prototype is developed to clearly define functional requirements. In other cases they are taken through and used as operational systems. This is particularly the case of small individual user-specific systems developed to run on personal computers.

The details of software development methodologies lie outside the scope of this book. However, some of the details, in particular those relating to

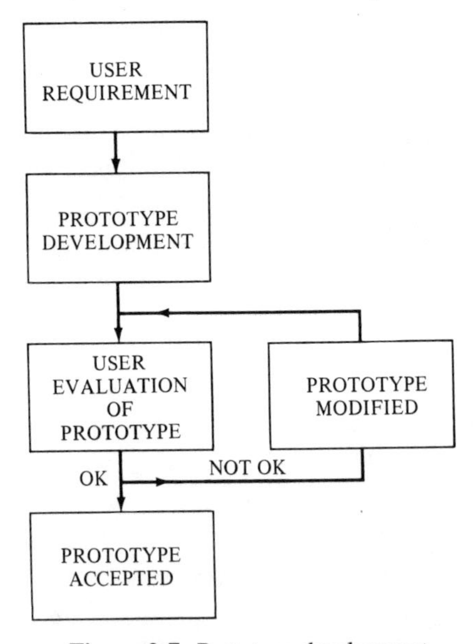

Figure 2.7. Prototype development.

modeling techniques used to simplify understanding of complex environments, are of interest. These techniques can be used at different levels. At the highest level, they can be used to describe the business environment. At a lower level, they can be used to describe individual functional systems, and at a still lower level, they can be used to describe individual programs. At a sufficiently detailed level, such techniques can provide for automatic generation of software. By linking the models at the different levels, the link can be made between strategy development and software development.

Depending on the characteristics of the environment being modeled, hierarchical decomposition, data-flow oriented, data-structure oriented, and control-flow oriented modeling techniques can be used.

The hierarchical decomposition and related control-structure techniques, focus on top-down decomposition of a system through its control organization. An example is the decomposition of a corporation into divisions, each of which is then decomposed into departments, then decomposed into groups, and so on (Figure 2.8).

Data flow methodologies help to analyze the flow of data through a system, show where it is processed, and where it is stored (Figure 2.9). An example is the U.S. Air Force's ICAM (Integrated Computer Aided Manufacturing) language IDEF. This can be used as a function modeling technique, decomposing each function into sub-functions, each of which is further decomposed into its sub-sub-functions, and so on. In each case, the basic activity, input, control, mechanism, and output are identified. This leads to a top-down hierarchy of data flow descriptions, with each level revealing increased detail.

Data-structure oriented methodologies, such as the Warnier methodology, aim to link the design and structure of a program to the structure of the requirements for which it exists. The methodology is based on the concept of an entity, such as a customer or a product, its attributes, and the data describing it (Figure 2.10). The complete business data structure shows how the attributes create relationships between the different entities in the business.

Control-flow oriented methodologies model a system or a function in another way. They describe the allowable values and combinations of the inputs and outputs to functions, and show how these are related to the detailed processes of the business. The PetriNet is an example of a control-flow representation (Figure 2.11).

These system modeling techniques have proved useful in developing

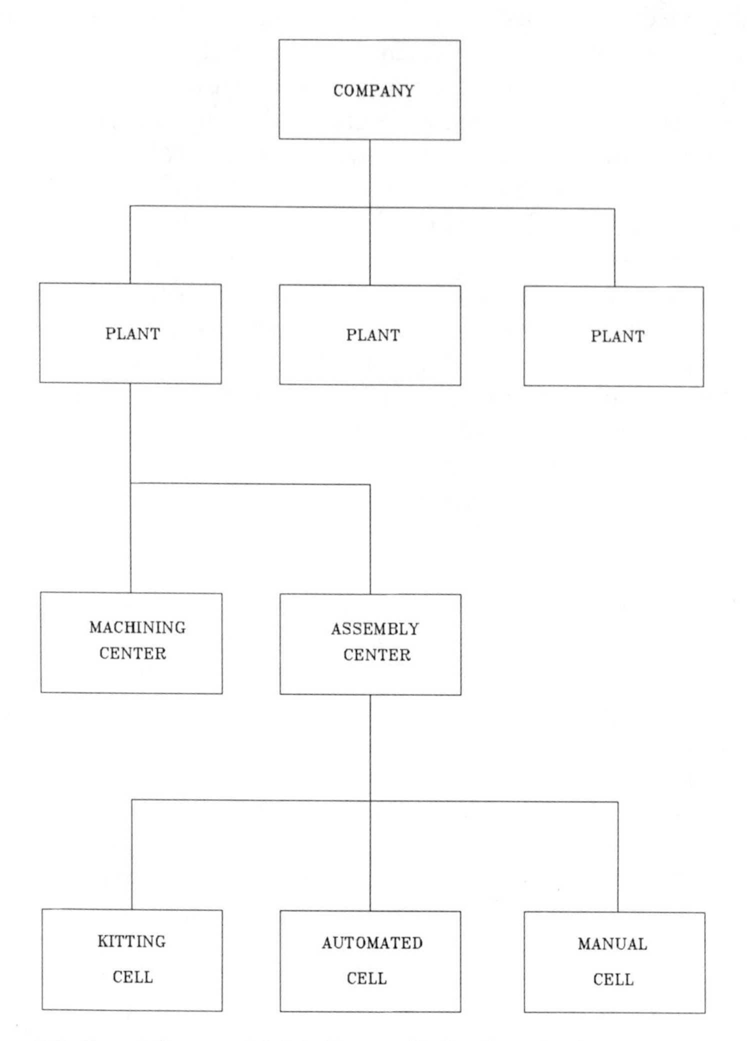

Figure 2.8. Control Structure Model. (*Source:* E. Gerelle and J. Stark. *Integrated Manufacturing Strategy, Planning, and Implementation*. NY: McGraw-Hill, 1988.)

software and systems. They are now often run on a computer, and can be used to model very complex systems, organizations, and even entire companies. They are easily understood and provide a basis on which managers, IT specialists and users can work together, and gain a common understanding of how a company works.

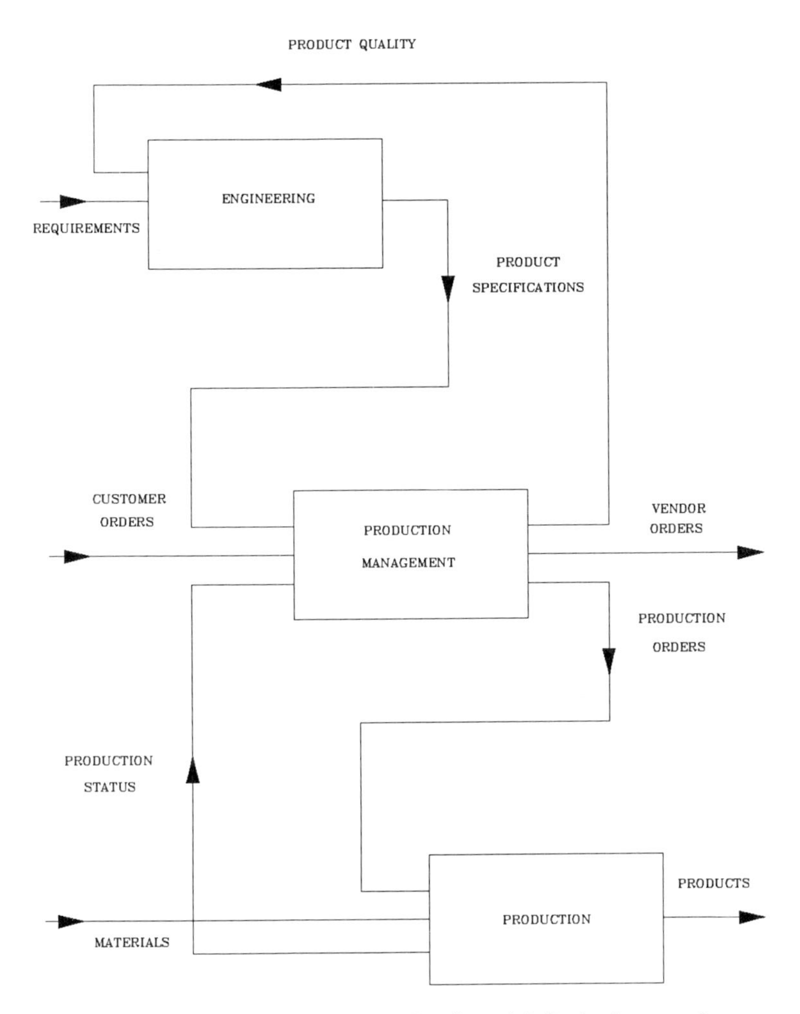

Figure 2.9. Data Flow Model. (*Source:* E. Gerelle and J. Stark. *Integrated Manufacturing Strategy, Planning, and Implementation*. NY: McGraw-Hill, 1988.)

2.5 INCLUDING IT IN THE BUSINESS MISSION AND STRATEGY

Hopefully, the reader who has gotten this far in the book will now believe, at least to some extent, that IT can be used to help gain competitive advantage.

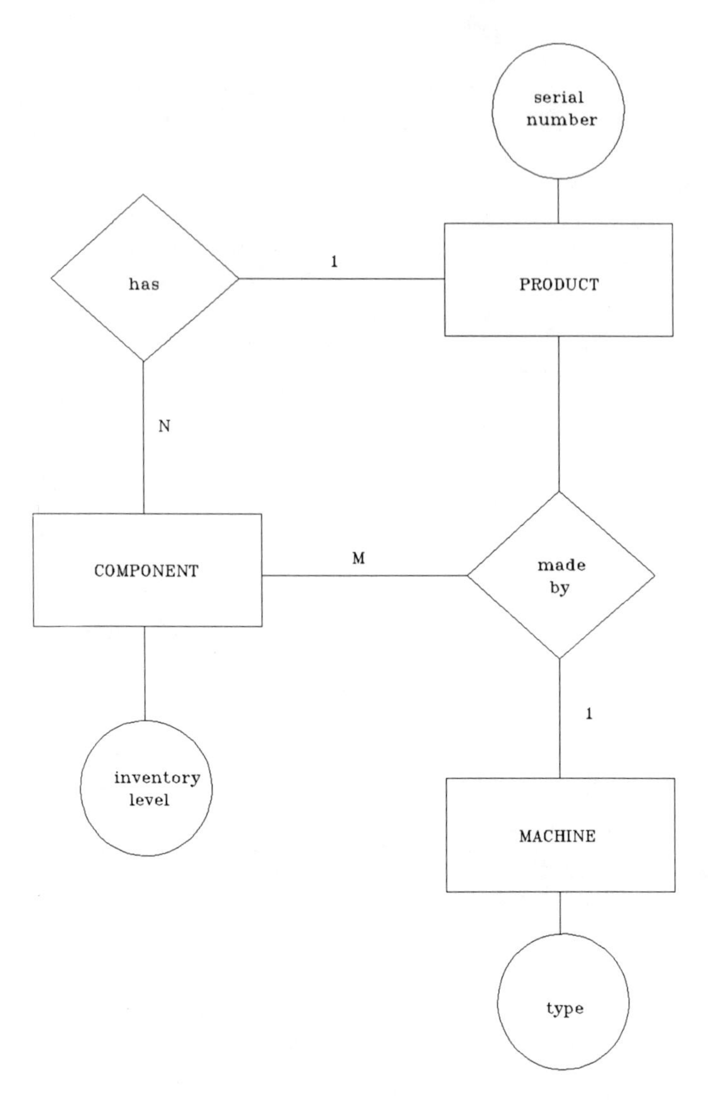

Figure 2.10. Data Structure Model. (*Source:* E. Gerelle and J. Stark. *Integrated Manufacturing Strategy, Planning, and Implementation*. NY: McGraw-Hill, 1988.)

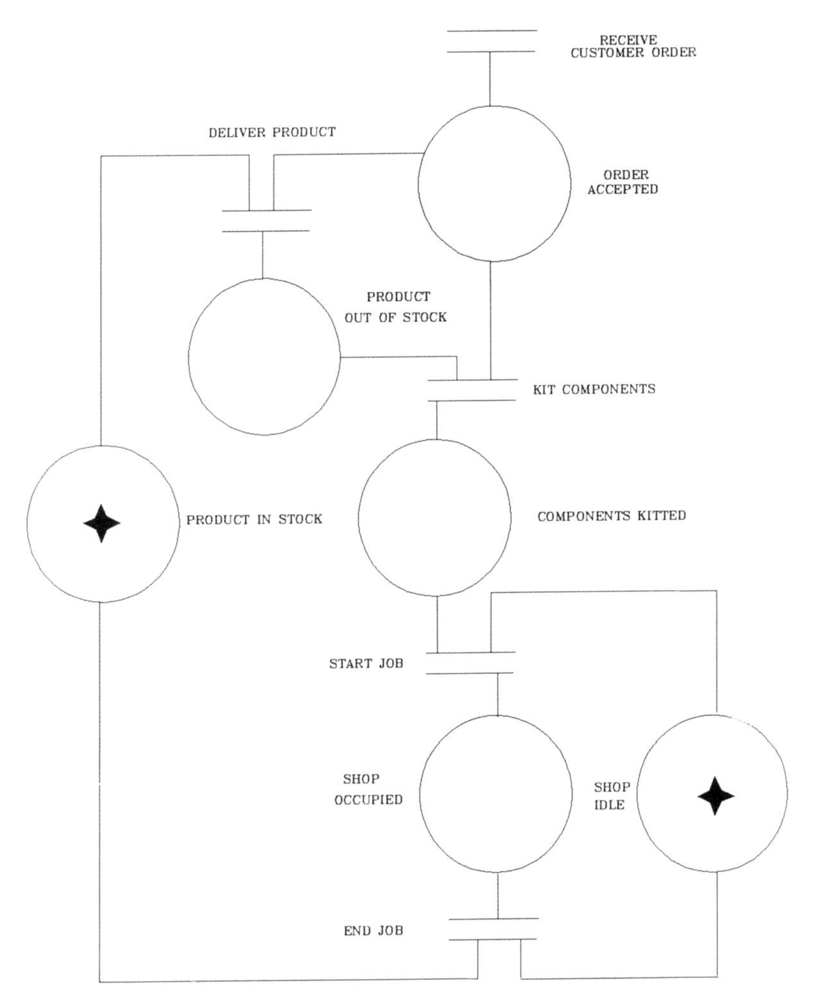

Figure 2.11. Control Flow Model. (*Source:* E. Gerelle and J. Stark. *Integrated Manufacturing Strategy, Planning, and Implementation*. NY: McGraw-Hill, 1988.)

However, having seen that the traditional business planning and EDP planning methodologies are not very useful in developing a suitable strategy, questions such as 'How do we do it?' and 'Where do we start?' may now be raised. The starting point will depend on the position of the company, but the steps to be carried out will always follow the same logic (Figure 2.12).

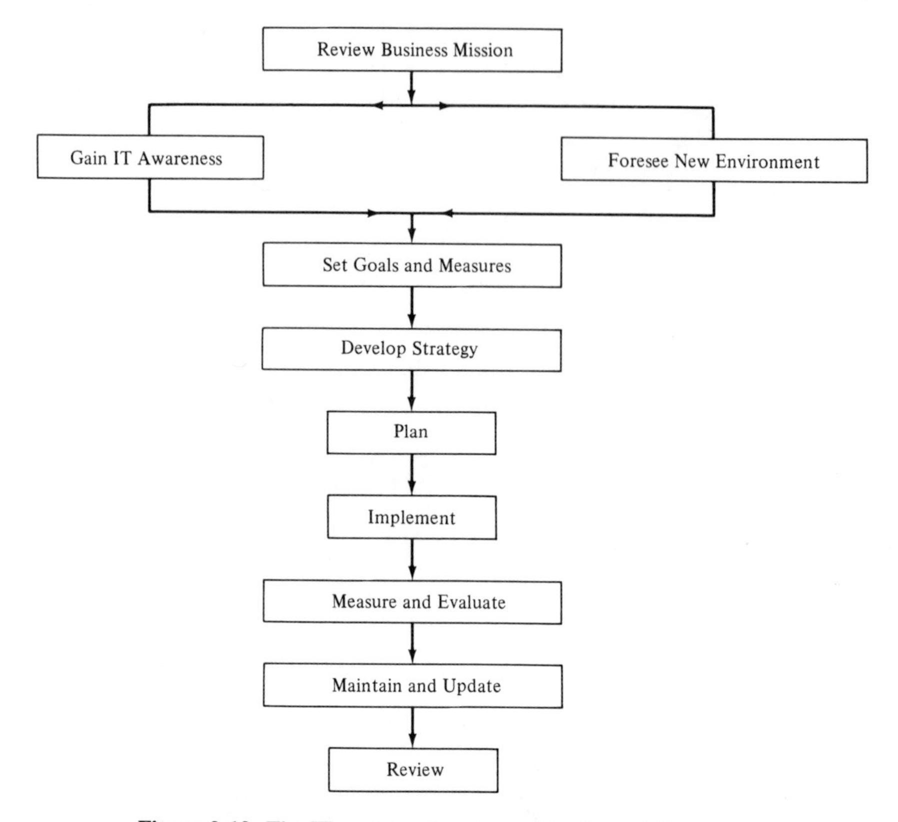

Figure 2.12. The IT strategy planning and implementation process.

The potential effect of IT on a manufacturing company is so great that it may lead to modifications in the business mission and goals and definitely must be considered when setting business objectives and defining business strategy. The IT strategy should be treated as an inseparable part of the overall business strategy, just as functional strategies, such as those for marketing and manufacturing are.

The business mission describes the purpose of the business. Such a statement should endure over a long period of time—say ten years or more. It serves as a foundation on which shorter-term decisions can be made.

Whereas the business mission gives a high-level and concise expression of the purpose of the business, definition of the business strategy leads to a much more detailed framework within which decisions can be taken and people can work. In particular, the business strategy shows how the company will act to meet the business mission and goals. The strategy takes into account short-term factors, such as current product strengths and weaknesses, customer perception of the company, and competitive behavior. When setting the business strategy, the top management team must make sure that it takes full account of the possibilities of IT, and that IT is considered a function as important as marketing, engineering, and manufacturing. If this is not done, then the strategic benefits of IT are lost, and the company will continue to drift along in the same undirected EDP way. The business strategy should be set looking three to five years ahead: it will need to be reviewed, but hopefully not seriously modified, annually.

The definition of the business mission and the business goals, and the setting of the business strategy are often described as if they occurred in sequence. In practice they tend to develop in parallel, at least initially, because when the mission is first set, the underlying details are not fully understood, and their definition may change some of the assumptions behind the mission.

There is a similar close link between setting business strategy and setting functional strategies; e.g., for manufacturing or sales. The functional strategies ultimately depend on the business strategy but at the same time are slightly independent in that there will be several possible functional strategies; e.g., for manufacturing, that will be consistent with the business strategy. The functional strategies will be set looking three to five years ahead. They may need to be reviewed annually and modified significantly as a result of local factors, which may not have a major effect on the overall business strategy.

Once the business and functional strategies are set showing how the company will meet its objectives, detailed plans should be implemented. The systems are put in place, procedures developed, and users trained. After implementation, top management should be looking to see that strategies are succeeding and objectives being met.

Periodically, the results of the strategy should be reviewed and compared to the original objectives. Due to the changing business environment, particularly customer and competitor behavior, changes may have to be made. However, unless the environment changes drastically or the strategy is obviously incorrect, major changes should not be required for several years.

To understand the results of particular strategies, it will be necessary to measure their impact. The measures taken should be agreed upon during the initial stages of defining the strategy: they may be quantitative or qualitative. In the absence of such measures, it will never be known if progress is being made with a certain strategy towards a given objective. (Knowledge of this type was almost nonexistent in the MIS—EDP environment, where such measures were rarely set or compared to performance.)

The measures should be related to the business goals, not to IT itself. Measures such as "3% of revenues spent on IT," "15% of capital budget spent on IT," "increase IT spend by 10% annually," "install 50 personal computers," and "20,000 hours to be worked at CAD screens" do not measure the success of the company's use of IT. More suitable measures would be "time to market reduced by 10%," "accounts receivable reduced by 20%," and "sales force making 20% more customer calls." These measures are closely related to the business goals. They must be set before the IT strategy is defined, not after. They are derived from the overall business strategy and are set by the top management team. If such measures are set, people will know what targets to work towards, management will be able to measure the process, and rewards can be linked to results.

Similarly, internal allocation of IT costs should be related to business goals, not to traditional EDP parameters, such as the amount of disk space used by a program.

Figure 2.12 shows the activities which should occur when a company develops an IT strategy to gain competitive advantage. It shows the links between business mission and goals, business strategy, functional strategy, planning, implementation, review, and measures of success.

With this background of business mission, business strategy, and functional strategy, the question will arise of where the use of IT to gain competitive advantage has to be addressed. The answer will depend on the age and activity of the company. A start-up consumer electronics company would want to address it in its initial definition of the business mission. A mature company in the power generation equipment sector might see no reason to immediately redefine its business mission. However, even such a company should consider IT at the very highest level. The three business objectives of marketing, innovation, and integration address the fundamental issues, and they provide the frame for everything that comes after. If IT is not taken into account when the objectives are addressed, then there is a danger that the company is ignoring the potential benefits of IT, and when the strategy is set, will not focus on using IT to gain competitive advantage. Can IT help the company better understand the needs of the customer? Can IT help the company to design, produce and deliver competitive products? Can IT help the company to supply competitive products under time and cost conditions that will be acceptable to customers? The answer to all three questions is surely positive. It is at the level of the business objectives that the company has to recognize that it can use IT to gain competitive advantage.

2.6 DEVELOPING AN IT STRATEGY FOR COMPETITIVE ADVANTAGE

The only way to develop an IT strategy that will lead to sustained competitive advantage is to include the IT strategy definition process in the business strategy development process. This implies that many of the typical business strategy development questions need to be extended to include an IT element. What volumes of products could be feasible if maximum advantage is taken of IT? How can IT affect the way products are sold? How can IT help to reduce overhead costs? How can IT help to improve the productivity of the Marketing function? Will IT be used to gain a particular competitive advantage, or will IT be used to support a strategy that the company has already developed to gain competitive advantage, such as being the lowest-cost source of a product? As these questions are asked it becomes apparent why they are so difficult for top managers to answer. The basic information needed to answer them is not available in most companies. Many companies are not even sure which questions are the most important ones to ask.

Figure 2.12 shows the overall process for developing and implementing the strategy. Figure 2.1 focuses on the classical four-step business strategy definition process. The first two steps involve the preparation of the basic information relating to the business objectives and the identification of the company's resources. To take account of IT, these two steps must be augmented by a creative analysis of IT that includes the following five issues:

- how the company works
- the areas in which IT can be used to gain competitive advantage
- the current state of IT in the company

- developments in IT
- competitive use of IT

The easiest and most useful way to understand how the company works is to develop a set of related models that reflect reality. This is an activity in which both IT managers and staff, and those from other functions, can participate and gain a joint understanding. The models should start with a high-level model of the business environment and extend down towards the more detailed description of individual functions and activities. To develop such models, people within the company must participate. As they learn from the model how the various parts of the company work, and are linked together, they develop a common understanding of the company. Their involvement at this time will help to get their support at implementation time.

The functions and activities of the company should be investigated from the point of view of both flow and use of information. Such an investigation should of course involve the people within the company who make use of this information. In particular, the middle managers and the current users of IT should be involved. These people know the company well and understand how information is and could be used along with the information requirements of their customers. It is often found that they have good ideas which are helpful in understanding how IT could be used more effectively in the future.

Those functions that are intrinsic to the company and set it apart from competitors must be understood along with the critical success factors for the business (discussed in Section 2.2 in terms of Quality, Productivity, Adaptability, and Flexibility). These will help to show in which areas the search for strategic application of IT should be focused. Similarly the key steps in the value chain must be identified. It is often in one of the areas that is of major importance to the company that IT can be used to gain competitive advantage. The key areas of the value chain that lie within the company must be investigated. How can IT be best used to support them, how can IT help to reduce time to market cycles, and would IT enable new functions and services to be packaged with the product; these and similar questions need to be asked with respect to the external parts of the overall value chain. How can the company use IT to improve the effectiveness of supplier activities? Would EDI be appropriate? How could IT help to bring the company closer to customers through marketing, sales, distribution, and service activities? Would common systems be appropriate? Such questions must be asked for the current situation and for the future.

What will the customer be wanting from the company in 5 year's time? It should be possible to work back from these needs to understand how the company should be reacting now. Modification of the position of the company in the value chain should be considered. Will IT enable the company to have a closer link to customers, reducing perhaps the role of independent distributors and maintenance organizations? Would it be effective for the company to try to reduce the customer's costs by taking over parts of the value chain and, incidentally, modify the competitive environment?

The current IT systems of the company must be understood. It must be made clear—in terms so that non-IT specialists can understand just what such systems should be doing, and how this meets strategic requirements. The following are suggested questions that should be asked: In which areas of the company is IT being used? Why not use it in other areas? What has been the experience of introducing IT systems, and what are current attitudes towards it? What is the IT competence level of the company and of specific individuals? Has IT only been introduced in the F&A activities, or is it also built into products? Has its use in different processes been integrated, is there integration between its use in products and processes? What proportion of current IT spending is directed to EDP maintenance activities, and what will be the long-term effects?

An IT scan should be carried out to understand what is available and what will become available in the next few years. As most technologies take several years to become freely available, those that can provide competitive advantage for the company in the short- or medium-term are probably either available on the market or close to coming to market. They may already be available in another country, or they may be under development in another industry, a university, or a defense program. The question must be asked—what new developments could help the company meet its goals? Also, as a result of expected developments, should the company be moving into new areas?

Although a technology that is widely available on the market may provide short-term competitive advantage, it alone in the absence of organizational actions, cannot maintain such an advantage for long as competitors are quick to follow suit. On the other hand, the wrong technology can easily lead to a company losing competitive advantage. Technology advances so quickly that it is possible for an inflexible strategy to cause severe problems in the long term. Consider a manufacturing company that decided in 1985, as part of its implementation plan, to base all IT activities on a centralized mainframe computer. Within a few years, such a company would suffer from not having Personal Computer, Engineering Workstations, and local communications networks.

The current and future IT behavior of competitors and potential competitors should be investigated. How are they using IT today (probable answer: don't know), and have they made their IT strategy public? Consideration should be given as to how a new competitor, without the company's historical skills and constraints would behave to make the most of IT. The IT behavior of suppliers should be understood, and their ability to be included in a common IT approach must be assessed. (If they have never used a computer, it may be difficult for them to imagine reading ordering information directly from the data base.) The customers' current and future IT requirements and perceptions must be understood. They may have a very advanced IT strategy, which the company will have to understand, or they may be aware of the IT plans of the company's competitors. The potential information requirements of other organizations affecting the market, such as regulatory bodies, must be understood.

None of these points can be covered in sufficient detail, if at all, in a classical strategy setting process. It is very difficult to carry them out without people who do not have a good understanding of the current and potential use of IT, and how IT can help to meet business requirements. Often it is necessary to get a specialist from outside the company to provide assistance. Once the company has completed the first two steps of the strategy definition process, it can move on to the other two steps. Each possible strategy must now be examined in detail, to show how it would lead the company towards meeting the business mission and goals. The particular competitive advantage that results from each potential strategy must be described, and the IT implications detailed.

For each strategy, the link to the strengths and weaknesses of the company and the opportunities and threats in the market, must be identified. Outline functional strategies, including those for IT, must be developed to show how activities would be carried out. The organizational structure, including the IT function, and the relationships between the various parts of the structure, should be reviewed. A financial appraisal of the strategy must be carried out based, not only on direct cost savings but also on the effect on market share and revenues of improved quality, productivity, adaptability, and flexibility.

The IT-related strategy that a company chooses will obviously be dependent on the company's business mission and the current situation. The strategy of a low-cost, market follower manufacturing company is very different to that of a state-of-the-art, high-functionality market leader. The role of IT will be a function of whether the company intends primarily to reduce cost as much as possible or seeks to differentiate itself in other ways from its competitors.

Although one may think that companies in emerging industries would want to focus on completely different areas from those that are mature or declining, it is rare to find such clearly defined cases. Even in declining industries, with commodity products, where the main differentiation should be cost, it is often found that IT can help the company to differentiate itself from the competition in other ways than cost, for example, in service. Although much of the demand may well be for mass-market, off-the-shelf delivery, there will also be a significant number of customized products, as well as a need to provide high-speed maintenance. The product/market groupings should be well segmented and individually understood. The potential for IT has to be understood for each one. IT may allow the company to reach different market segments through different channels. In some cases it is possible to gain an advantage by automating routine activities or using an expert system to help in a very complex activity. In other cases IT may provide the adaptability and flexibility required to address a niche market.

Some companies may use IT to change their business mission becoming, for example, information companies rather than manufacturers. Some will see the potential for introducing new products, such as video games, graphics tablets, and satellite antenna, which result from developments in IT and communications. There will be companies who focus their use of IT on adaptability, rapidly enhancing their products to remain market leaders and perhaps modifying the value chain at the same time. Others will concentrate IT on increasing productivity by driving down costs. Many will use IT to tackle the twin objectives of flexibility and adaptability, simultaneously increasing the rate of new product introduction and decreasing product development time.

The company may use IT as the basis for bringing together people from different functions or different locations to work on the development of new products. IT may provide the backbone for a total supply chain perspective that can give a faster response to market changes. Also it may present the company with an opportunity to remove a barrier that a competitor has put in place, or create a barrier that will make it more difficult for a competitor to reach a customer.

The major force behind the drive for an IT-related strategy may be a customer, used to working with product teams, who wants to include company staff in these teams and also be able to read company inventory levels on a day-to-day basis. In such a case the company is obviously going to have to focus attention on its communications capabilities and the compatibility of its systems with those of its customer. Depending on customer requirements, another possibility could be to use IT to master the most complex part of the value added chain and become the best player on the field in this particular area.

The results of the IT strategy development process should be made available in the form of a concise report. This should explain why a certain course needs to be taken, what this involves and, at a very superficial level, how it will be implemented. It should contain an overview of the architecture—the overall layout of all the IT (computing and communication) resources. It should also show what hardware will be used, where it will be used, and how the components will be connected. Similar overviews should be available for the software, the data, and the human resources. The role of IT within each function will be addressed, and the expected results described. The organization of IT resources will be defined, showing for example the structure of the IT function itself, and the way it will work with the other functions. Any particularly important issues, such as the use of international standards, policies for recruitment and training, and methodologies to be used will also be addressed.

2.7 ORGANIZING THE COMPANY FOR COMPETITIVE USE OF IT

The subject of this section is the overall organization of the company as a function of the IT-related business strategy. There are two aspects to this subject: the first involves the role and position of the IT function within the company; the second involves the effect of IT on the organization in general. A related subject, the organization of the IT function within a company oriented towards using IT for competitive advantage, is discussed in Section 2.8.

The organization of the company is directly related to the chosen business strategy, the organization should not be identical in any two companies, because no two companies will have identical business strategies. Nevertheless, there are several organizational points common to most companies which intend to use IT to gain competitive advantage.

The organization has to be defined by the top management team. It is of course company-wide and so cannot be addressed from within a single function. In particular, even in an IT-oriented company, the overall company organization cannot be defined by the IT function alone.

Overall responsibility for use of IT and communications within all the functions of the company should be given to one person. This person can be referred to as the IT VP (Information Technology Vice-President). The IT VP must be a member of the top management team. The role and responsibilities of this person are described in the next section where it is shown how they are very different from those of the traditional EDP Manager. The IT VP is the leader of the IT function within the company and reports directly to the CEO. He/she does not report to either the VP Finance or VP Administration and should not be subservient to any individual management function within the company. The IT VP will be responsible for making sure that IT is used efficiently, both within the company and in its relations with the outside world.

The organization should be developed so that resources are used as well as possible to meet the aims set by the business strategy. Depending on the strategy chosen, the organization may have a strong internal focus, or it may be directed outwards towards markets or particular clients. The internal focus may be vertical, addressed at one of the major functions, such as manufacturing, R&D, engineering, F&A, or sales and marketing. Alternatively, it may be more horizontal with the intent of bringing several functions together. Whatever strategy has been chosen, the whole company must be organized to respond to it. In the same way that no strategy will completely ignore any one of the functions of the company, no organization should completely ignore any of the functions. (This is very different to the traditional IT organizational approach, which has often seen IT subservient to the F&A function, buying a mainframe computer to provide service to F&A, and then refusing to provide service to R&D and the Shop Floor because these functions have no use for the mainframe computer.)

Some of the IT-related resources will go to the IT function, which in most manufacturing companies will be a support function. Other IT-related resources will go directly to those functions that provide competitive advantage. The IT investment will be distributed in the direction chosen by top management and expressed in the business strategy and not allocated by an EDP Manager working for the F&A Department. In most cases the IT investment will not be made, as in the traditional organization—80% to F&A and EDP and 20% to the other functions. There will be some cases, of course, where this distribution actually reflects the competitive requirements, but in most cases, F&A will receive less than 20% of the resources, and the Manufacturing and Engineering functions in particular will receive significantly more.

In the traditional EDP organization, the IT function reporting to F&A has often been given too much power over the experts in other functions. It should be remembered that in most cases it is the other functions that really provide the competitive advantage, and IT only offers a way of supporting them.

As the organizational structure depends on the business strategy, it will not be the same for all companies. There will be few cases where it will have its traditional aspect (Figure 2.13). At the least, such a picture should make way for the IT function (Figure 2.14), reflecting the fact that the IT VP reports directly to the CEO.

Related to the traditional departmental view of the company is the hierarchical control pyramid (Figure 2.15). This too will change because it does little to provide competitive advantage but much to prevent free flow of resources and information. It prolongs the belief that all control and information flow within a company is vertical. In practice this is not the case. The best illustration of this is perhaps the "product team" in which

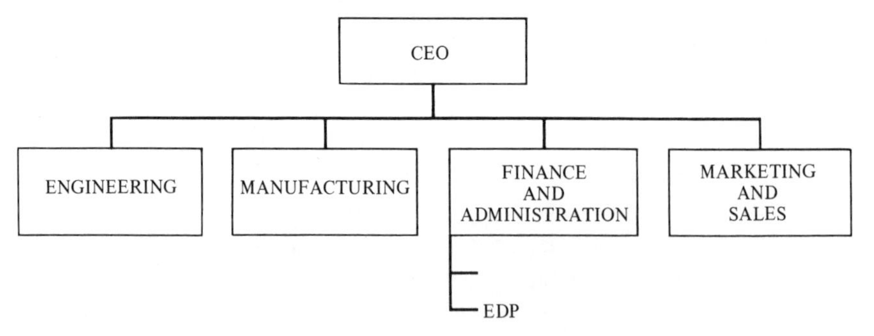

Figure 2.13. The position of EDP in the traditional organization.

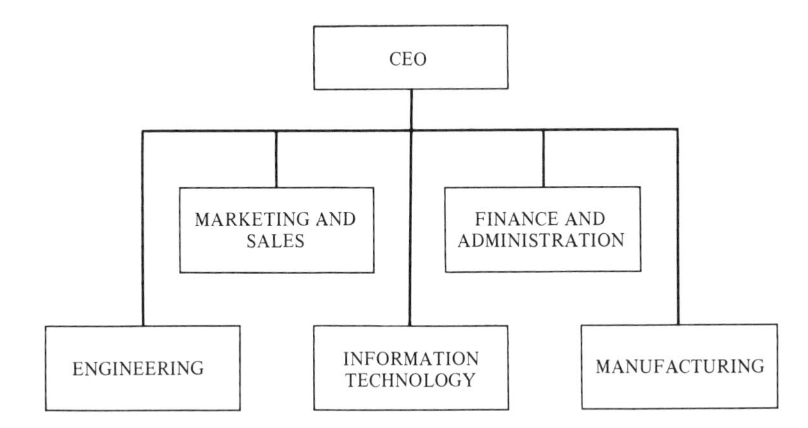

Figure 2.14. The IT function reporting directly to the CEO.

individuals are drawn from different functional areas, such as marketing, engineering, manufacturing, finance, and after-sales service, to bring together their knowledge and experience to provide the customer with the best product at the right time. (The designer of a part and the person who machines it should communicate directly, not through ten levels of hierarchy.) Even in such an organization, most of the information does not flow across functional boundaries but remains within a given specialty area.

From an information point of view, the organization must reflect the fact that most market-, product-, and process-related information is used and transferred within a function and transferred horizontally to other functions. Only a small fraction of the information, in addition to the control information, flows vertically. The small amount of information flowing vertically leads, of course, to questions about the many layers of middle management found in most companies. In the past many of these have done little more than relay information from one part of the company to another. As many companies have realized even before implementing IT to any great extent, the number of layers can be reduced significantly: with increasing use of IT, the number of layers of middle management will decrease even further. At the same time, the improved communications available with IT will increase the number of subordinates reporting directly to one person. The result of these two forces will be the creation of flatter, decentralized organizations recognizing a larger number of specialist functions that can be brought together and focused on particular products and problems.

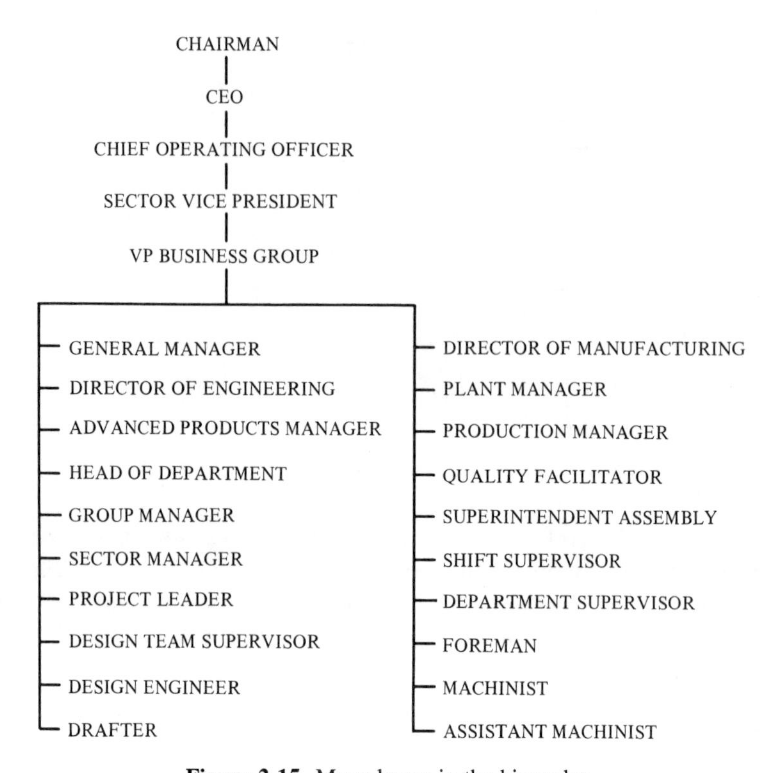

Figure 2.15. Many layers in the hierarchy.

Most information in a company stays within the same activity: most flows horizontally, a little flows vertically. Changes may be needed in the organization to make sure that the horizontal flow of information is as effective as possible. In the past, department barriers have often led to unnecessary activities being carried out with information and to a demarcation of activities between departments that was irrelevant in the IT-oriented company. As a company begins to use IT to gain competitive advantage, it may well be necessary to redefine functional boundaries, redistribute activities among the functions, and reduce administrative overhead on information transfer.

Another problem with the traditional departmental representations of the organization is that they take no account of the external environment. From the IT point of view, Figure 2.14 is a useful first step towards presenting a truer picture. Although Figure 2.15 may help the CEO feel better, it may be far from the truth, with the real power lying with the suppliers or the customers. The organization should take into account the competitive environment that surrounds the company. Interfaces with the external world are areas that are critical to success and are strong potential candidates for competitive use of IT.

Compared to traditional EDP organizations, which have often been called "centralized" because most of the company's computing was carried out on a central computer, the use of IT resources in future will appear decentralized and dispersed. However, as these terms can be very misleading, more explanation is needed. The IT control and command structure should be "centralized" in the person of the IT–VP, to act in the direction set by the top management team. The opposite of this is not "decentralized" but "anarchy." The computing resources do not need to be centralized. They can, and should, be decentralized so that they are available for those who need to use them. Similarly, the responsibility for their use should be with those who use them. Thus in the Engineering function, major systems like CADCAM (Computer Aided Design Computer Aided Manufacturing) and Engineering Data Management should be under the control of, for example, an Engineering Computing Systems Manager backed up by experts who understand these systems, and the way they are used. This is very different from the situation in the traditional MIS—EDP environment. (There are very few traditional EDP Managers/ MIS Directors who have any understanding of functional systems beyond the F&A area.)

In the same way that Engineering may be seen in a given company as a very important competitive function, other companies may also see Marketing, Sales, Manufacturing Planning, or the Shop Floor as areas where they can use IT to gain competitive advantage. In each case, there should be a functional Computer Systems Manager, who belongs to that function and understands its requirements and takes responsibility for implementing IT within this area to meet the requirements of the top management team. This person should have available a team that will include IT specialists and users of the system with good IT understanding. The functional systems staff should learn about the overall functional process and understand it at least as well as users who have two year's experience. The functional Computer Systems Manager will report to the VP of the function (Figure 2.16).

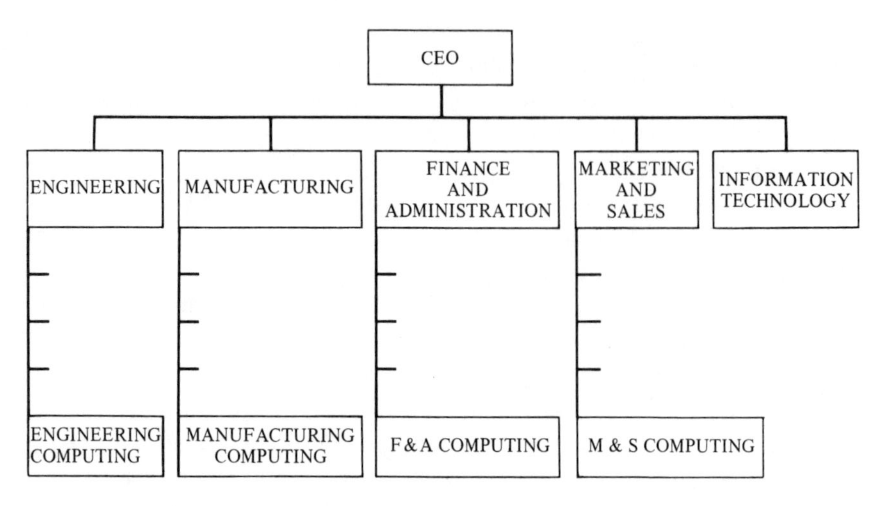

Figure 2.16. Functional Computer Systems Management.

2.8 THE IT FUNCTION SUPPORTING COMPETITIVE USE OF IT

In most manufacturing companies, IT is not the key function providing competitive advantage. IT will mainly be applied to another function (or functions) so that this function becomes and stays competitive. The IT function should be a support function. The major objective is to ensure that IT is used as specified by the top management team to gain competitive advantage.

The IT function should be led by the IT VP. This person will be a member of the top management team, reporting directly to the CEO and will have a prime role in ensuring that the business strategy takes full account of the possibilities offered by IT. The IT VP will be responsible for all the computers and networks used by the company and for all IT use across the company and with the external environment. He/she must ensure that service is supplied to all the functions of the business as specified in the business strategy. Apart from participating in the development of the business strategy, the IT VP will be responsible for developing a detailed IT plan, implementing the IT-related parts of the business strategy, and for setting the corresponding information and IT policies. The IT VP will have to tie IT goals and rewards to business success, and be responsible for the communication between business managers and IT specialists. The IT VP is responsible for short-term systems performance but also for meeting long-term requirements, which will surely impact the organizational structure.

The IT VP should keep in close contact with progress on implementation of the IT strategy, and report frequently on its progress to the CEO. As IT will now be a major component of the business strategy, it should receive due attention from the top management team, but if this does not happen the IT VP will have to find ways to make sure that it receives its fair share of management time and support.

To understand the IT function in the company, it is helpful to compare it to the traditional EDP function. Typically, this function reports directly to the F&A Department, and often has no relationship with other functions. It operates a mainframe computer, offering the standard packages available from mainframe vendors. When company specific software was required, it got involved in resource-consuming activities of analysis, programming, and maintenance. Often the end result will not have met requirements. The EDP function will have been heavily influenced by the mainframe vendor: when the mainframe vendor introduced MIS systems, it implemented an MIS system. When Personal Computers were offered by the vendor, it will have bought Personal Computers and when electronic communications were offered, it will have designed a network.

Typically it will act more in response to the new products made available by the mainframe vendor, than to business requirements. This has led to the user view that EDP has been too slow, too expensive, too late, too complicated, and out-of-touch. Management's view of an EDP Manager is of someone too technical, speaking in jargon, with little understanding of strategy, and not a potential member of the top management team. The EDP Manager prefers low-risk, low pay-off projects, which try to preserve or improve the image of MIS—EDP, rather than embarking on more essential, but more risky, projects with potential high pay-off for the company. The EDP Manager will probably never think about using IT to gain competitive advantage.

The IT function's only reason for existence is to respond to business requirements. Its role is to support other functions, not to try to dominate them, or to try to set itself up as an independent nonproductive empire. If the IT function cannot, for example, help implement the Shop Floor Control systems required to attain competitive advantage, the responsibility lies with IT, not with the Shop Floor. The IT function has to provide a service to the other functions within the company. If it cannot do this, then top management will have to take the appropriate action, which may involve changing the IT VP.

To clearly understand the work that the IT function has to do, it is useful to compare it to the work of a manufacturing company. The organization chart of such a company (Figure 2.17) shows that there is a role to develop strategy, direct and manage the company. There are specialists who carry out research into new products and process and those who develop new products and processes. There are some who make products, some who look for areas where clients have needs, some who sell products, and others who provide after-sales services.

There are many similarities between these activities and the activities that are associated with the IT function. The traditional EDP function has not filled all of the roles. Often it has only filled those that were of interest to EDP mainframe oriented specialists. Its first love has been to operate the mainframe, "tuning" it for better performance, and in some extreme cases, modifying the operating system software. This type of activity, operation of the production equipment, is a normal activity of the manufacturing function of a company.

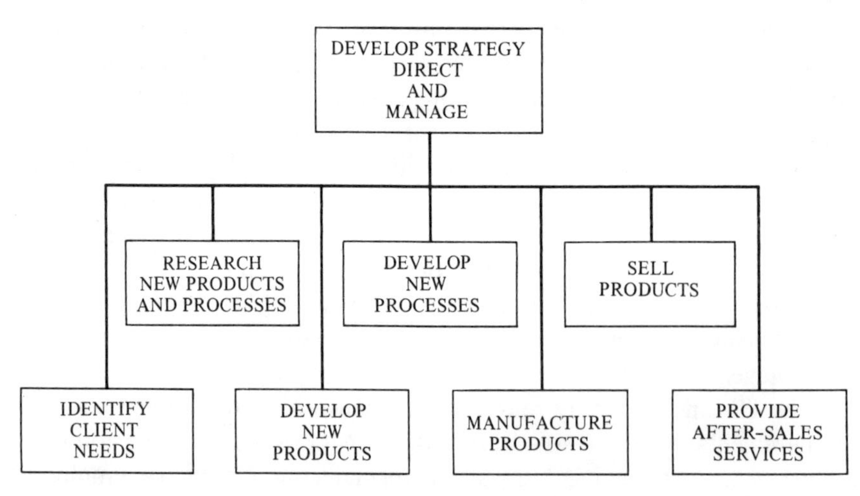

Figure 2.17. Typical activities of a manufacturing company and of the IT function.

Later, traumatized by the invasion of microcomputers and the decentralization of the EDP resources, the EDP Department has involved itself in providing the basic word processing equipment and communication networks but rarely has it understood the possibility for the strategic use of these. Its role here has, once again, been that of providing the machinery which others will work with.

The EDP function has also been involved with other activities.

It has analyzed requirements and developed programs. This can be equated to the process development activity in a manufacturing company, since programs are the "processes" that drive the EDP "machinery." (The real "product" of IT being the business use of the programs.) It has kept abreast of EDP methodologies and other techniques. This can be equated to a work methods and organization activity. It has kept an eye on the development of new technologies in computing and communications much the same way as manufacturing engineers follow trends in manufacturing technology.

The activities mentioned so far are similar to those found in manufacturing operations and engineering. Had EDP restricted itself to these activities it might have been more successful.

It has however also involved itself in strategy (deciding what type of support it should offer different parts of the company) and after-sales support (of users). Its involvement in strategy has been unsuccessful because it has largely ignored the business environment. Its after-sales behavior has been unsuccessful as it has largely forgotten its internal customers (disparagingly called "users") and the company's customers.

In the one function, there has been an attempt, perhaps unwittingly, by the EDP Department, to get involved in a very wide range of very different activities, and often the same people have tried to do several very different jobs. Not surprisingly, in most traditional organizations, the EDP Department has failed in some of these activities. This is not to say that it should not have been involved in them. The IT function in the organization striving for competitive advantage must be involved in these activities and be successful with them. It must however be aware that it is involved with them, and be prepared to manage each one differently. It must be understood for example that an IT specialist who is good at organising the mainframe operators on a day-to-day basis will not necessarily be the best person to head up the development of a new and highly sophisticated piece of software.

The way that the individual activities will be grouped into sub-functions will change from company to company, but a typical example could be the following:

- IT Strategy
- IT Development
- IT Operations
- IT Service

The activities of the IT Strategy sub-function would include understanding the company's business strategy, the competitive environment and customer's requirements, and identifying possible IT products to help the company's business performance. The use and flow of information within the company would have to be clearly understood, as would its value and expected future use.

The IT Development sub-function is expected to follow technology trends and select technological solutions. It would also be responsible for systems planning, design, development, implementation, and integration.

The IT Operations sub-function is expected to operate and maintain the company's mainframe computer, the company's communications network, and any other computers and communications for which it was given responsibility. These functions are not strategic and may well be moved out-of-house, so that the company can focus on activities that contribute directly to business success.

The IT Service sub-function is expected to be in constant contact with IT users throughout the company, understanding their requirements, training them to master IT, and helping them to use IT.

Within the other functions of the company, such as Marketing, Engineering, Manufacturing, F&A, and Maintenance, the responsibilities for the above IT Development, Operations and Service sub-functions could either be assumed within the function, sub-contracted to the IT function, or shared with the IT function. For example, Engineering might want to be heavily involved with the IT Development and IT Services sub-functions relating to the Engineering function, but sub-contract Operations of the turnkey CADCAM system and the mainframe Engineering Data Management System to the IT function.

In the short term, the person who will perhaps be most affected by the change to using IT for competitive advantage will be the EDP Manager. If this person is appointed as IT VP then the problem will be solved for the short-term, only perhaps to reappear later if the person is not suitable for this role. In many instances though, it will be clear that the EDP Manager will not make a suitable IT VP. In this case two problems arise—what to do with the EDP Manager, and where to recruit an IT VP. The EDP Manager may choose to stay, taking responsibility for one or more of the IT sub-functions. The IT VP will not be easy to recruit because this person will be in demand from many companies, and there are few such people available. They must have both a good business knowledge and a deep understanding of IT. Additionally, they should be acceptable to the top management team, in particular to the CEO, who will entrust this person with a role that is crucial to the company's future.

3

INDUSTRIAL EXAMPLES OF COMPETITIVE USE OF IT

3.1 INDUSTRY-WIDE LESSONS

The third part of the book gives examples of companies obtaining competitive advantage from using IT. Not all of the examples are from manufacturing companies. Examples from other sectors can be useful in meeting the objective of this book and putting top managers in the position where they can use IT to gain competitive advantage. Getting to such a position requires an understanding of how IT can be used to gain competitive advantage. Part of this understanding comes from examining how companies in very different sectors, in particular the service sector, use IT.

The most visible and publicized examples of using Information Technology are found in service industries, such as banking, insurance, and airlines. Although these areas are outside the scope of this book, it can be useful to study them because many of the lessons learned in service sectors can be applied in manufacturing industries. Most readers will have practical first-hand experience as a 'customer' of these systems. This should help them to relate to the techniques described.

Examples from distribution and retailing should not be considered out of place, given the manufacturing context of the book. Rather they represent an important part of the manufacturing company's value chain. As such they are equally valid target areas for the application of IT to gain competitive advantage.

The reader should not look at the first examples and assume the others will be repetitious and not worth reading. Only by reading examples addressing different industries and different parts of the value chain, will the top manager be able to develop from the, sometimes common, experience of many industries, an understanding of the type of approaches that will be meaningful in one's own company.

Some of the most interesting and successful examples of competitive use of IT are associated with airlines. Airlines were one of the first major users of IT. Originally they used Information Technology in a conventional EDP way to manage the stock of seats, to reserve seats, and to issue tickets. Later, travel agencies were supplied with terminals so they could make reservations for their customers. This opened up the possibility that the travel agencies would also become retail outlets for the airlines to sell other services, such as hotel rooms and rental cars.

By closely studying customer behavior, the airlines found that the majority of passengers would choose a flight from the first list that came up on the reservation screen. More sales would result if an airline's own flights appeared on the first list, and flights of other airlines were on later lists. Those airlines able to implement this technique on their reservation systems increased market share at the expense of competitors who did not have the right systems and infrastructure. The very existence of such systems was a barrier to market entry by potential competitors. For a time, the reservation systems of some airlines dominated the market. Eventually this type of practice was banned, but while it lasted, it was highly successful for those airlines in the position to exploit the possibilities.

Airlines use their systems for yield management—juggling the number of seats available at different prices to maximize revenues. The better the system, the easier it is to judge the right mix and the right time to change the parameters.

Airlines have also used IT to build up their "Frequent Flyer" programs. Initially these were aimed at increasing customer loyalty. The next step, of linking up with a credit card supplier so that purchases made by a passenger with a particular credit card would count towards Frequent Flyer Awards, increased loyalty, was good for the credit card issuer, and brought in valuable information on passengers' purchasing habits.

Airlines are now looking to link their reservation systems with office automation systems so that secretaries will be able to reserve business flights directly from their desk. To improve customer service, airlines have equipped workers with computer terminals so that they can track down delayed and lost luggage, set corrective action in motion, and keep passengers informed of what is happening. To avoid flying planes full of empty seats reserved for passengers who have not shown up, the airlines overbook flights. The amount of overbooking acceptable for a flight depends on a variety of factors, such as the locations served, the month, the day of the week, and the time of day. Sophisticated forecasting techniques are applied to ensure, on one hand that planes are as full as possible, and on the other hand that passengers who do show up, get a seat.

Although the operations of an airline may seem very different from those of the typical manufacturing company, there are similarities from the point of view of using IT for competitive advantage. The reservation of seats can be compared to order taking. Forecasting the amount of overbooking allowed on a given flight on a given day has parallels with forecasting the needs of a given bakery on a given day. Frequent Flyer Awards, geared to gaining passenger loyalty, translate into supplier and customer loyalty schemes. Many service industry activities are information-intensive, close to the customer, and involve little material handling. Where these types of activity are found in manufacturing companies; e.g., in marketing and sales, IT can play a similarly competitive role. Likewise, a comparison can be drawn between the IT-assisted 'back-office' activities of service industries and similar activities in manufacturing companies.

3.2 AEROSPACE AND AUTOMOTIVE

In the aerospace industry, one manufacturer has invested heavily in aerodynamic and structural analysis systems to become a leader in the design of aircraft wings. Advanced computer-controlled wing production facilities make this manufacturer a leading candidate in the design and manufacture of wings. A similar investment by another manufacturer led to a strong position in military aircraft design. A third company used IT to help gain a leading position in the aircraft engine market. A helicopter manufacturer developed an integrated system focused on the aerodynamic analysis, sculptured surface geometric modeling, and NC machining of rotor blades. The result was a new blade that allowed one of the company's helicopters to establish a world speed record.

CAD and other computer-aided engineering techniques have been used extensively in the four-nation European development of the Airbus family of aircraft. For the A320 all 200,000 component drawings produced by some 2,000 engineers and drafters in Germany were produced on CAD, none by hand. The Airbus helped reduce the U.S. share of the world civil aircraft market from about 95% in the early 1970s to about 80% in the mid-1980s. All indications are that this decline will continue into the 1990s.

Many aircraft are now equipped with on-board computers that continuously monitor aircraft performance. Information can be processed initially on-board, and if necessary, transmitted to base where analysts will decide on the action to be taken. In most cases the analysis will show the need for standard or early maintenance, but in some cases action by the flight crew may be required, even perhaps an emergency landing. Manufacturers unable to offer such systems, are clearly at a disadvantage. Other on-board computers are at the heart of fly-by-wire control systems and can automatically help to protect the aircraft against dangers, such as stalling and windshear.

In the automotive industry, some manufacturers have used IT to reduce the time cycle from initial design to getting a car on the road from 5 to 7 years to 3 years. By the end of the 1990s, Japanese manufacturers expect to have this cycle time down to 1 year. Their competitors must either find ways of matching such performance or go out of business. IT can help by speeding up design work and managing the vast amount of data generated during the design cycle, and designs can be kept open until the last possible moment so that all the latest developments can be included. Structural analysis can be used extensively in weight reduction and in improving security in crashes.

Automotive suppliers have been pressured by manufacturers into switching from manual order processing systems to computer based methods. Although this has been a difficult and major step for them, it has resulted in inventory reduction and cash flow improvement. Just as importantly, it has changed the role of sales representatives from being order-takers to a more marketing oriented function that requires them to understand customer needs. It is estimated that further moves to completely replace paper transmission of order, invoice, and notification information by electronic transmission will result in savings of several hundred dollars per car.

Following the acceptance of computer-based order processing systems and the electronic transmission of orders between car manufacturers and their suppliers, the next step will be the transmission of design data by similar means: this should help to reduce lead times. From a manufacturer's point of view, the ability of a supplier to work in this way will be an important prerequisite for suppliers participating in the manufacturer's 'simultaneous engineering' activities. Suppliers that do not make the necessary investment will find themselves at a competitive disadvantage.

Some automotive manufacturers now supply their dealers with up-to-date information on special promotions, prices, conditions, discounts, options, colors, and technical and financial information. As a result, dealers can sit down with potential buyers, find out what the buyer requires, and produce a personalized document showing the exact specifications of the car and a personalized financing plan. The system can help customers understand the details of different loan deals and provide preliminary acceptance indications by the loan institution. Dealers can see current stocks on a screen, rather than waiting for information to arrive by post—three days out-of-date. Integrated information systems improve the car manufacturer—dealer relationship, giving dealers better control over their business and manufacturers rapid feedback on customer behavior.

Car manufacturers send out computer disks to well targeted customer audiences. The disks contain a wide range of information on either selected models or even the manufacturer's complete model line-up. Potential customers can check out specifications at home and compare the cost of different options and financing packages.

Some car manufacturers have supplied dealers with computer-based training packages to help train mechanics to understand the complexities of new models. This will facilitate the learning process for the mechanic, and in the long run improve the service to the end customer, in this case the car owner.

3.3 CHEMICALS AND PHARMACEUTICALS

Plastics manufacturers have developed on-line databases that can be consulted from customer terminals for details of the properties and prices of available plastics. Customers without on-line terminals are regularly sent updated diskettes that can be used on standard Personal Computers.

Even for an apparently low-technology industry as agriculture, IT can be used to gain competitive advantage. Mathematical models are built to describe agricultural crop/environment interactions. On-line information from sensors can then be used to propose the most suitable spraying/fertilizer procedures.

In another example, instead of buying the same fertilizers as everyone else, a farmer can take soil samples in to be analyzed. The best fertilizer mix is identified, and the farmer buys the most suitable product, rather than the bulk product that the manufacturer decided to produce.

A supplier of chemicals data sheets switched from paper files to a computer-based system. Response time for customers went down from over 5 minutes to less than half a minute.

In the pharmaceutical industry, where R&D spending can run over 10% of revenues, yet still grow annually at 30%, new systems have been introduced to enable R&D and Marketing to work more closely together, designing clinical studies that will show where market needs exist and how new products can best be positioned. Information systems costing less than $500,000 have been introduced in companies with more than 100 R&D staff and have led to 50% productivity increases.

It can take more than $100 million and 10 years to develop a new drug that will only have a 15-year life. To pay for the massive R&D investment, big sales are needed. This implies improved R&D and marketing, bigger and better sales forces, and probably consolidation of the industry. Different companies respond in different ways, but an attempt to use IT to gain competitive advantage is usually high on the list of actions to be taken.

Before a drug can go on the market it has to be approved by the FDA. This can take up to 3 years, as several hundred thousand pages of reports, graphs, and test results are checked.

Pharmaceutical manufacturers use image processing techniques that allow them to get all this documentation onto an optical disk, which can then be used by the FDA along with the physical documentation. It is estimated that this could help them get their product to market one year earlier. Going one step further, one manufacturer installed terminals, connected to its in-house testing and documentation system, in FDA offices.

Some pharmaceutical companies have installed terminals in hospital purchasing departments, linked into their own sales and inventory systems. Both sides benefited from such a situation. The hospitals gained through better coordination of purchasing and inventory activities. The pharmaceutical supplier gained a privileged relationship with the hospital, which competitors found difficult to overcome.

A drug wholesaler gave pharmacists terminals so that they could order directly through the wholesaler's stock and distribution system. The pharmacists benefited from more timely ordering and delivery. They were also able to use analysis programs, supplied by the wholesaler, to track behavior of their customers. The suppliers of drugs to the wholesaler were also interested in the information in the system. This led to the implementation of Electronic Data Interchange between the wholesaler and its suppliers, resulting in reduced purchasing and stock costs for the wholesaler.

A manufacturer of personal hygiene products used IT to help handle the problems of distributing a range of over 500 products to several hundred thousand retailers, ranging in size from corner stores to big-city supermarkets. Central computers and computers in the warehouses keep tabs on stocks, sales, and orders and have helped reduce stocks from 5 to 3 weeks. Trucks are loaded for maximum efficiency. Point-of-sales systems in the larger stores track consumer preferences and are used in predicting demand. New developments concentrate on the small retailers, helping them to reduce their stocks and to better match the range of products they carry to the local customer.

3.4 ELECTRONIC AND ELECTRICAL INDUSTRIES

The electronics industry has been at the forefront of using IT systems to gain competitive advantage. Typical results in manufacturing show build times halved, batch sizes drastically reduced, WIP and finished goods inventory slashed, inventory turns increased, and quality rates near 100%. The percentage of on-time orders has risen dramatically. Warranties have been extended and maintenance charges reduced.

The design time of new circuits has been reduced while their complexity has increased. Computer-aided simulation has cut development time by 75%. Use of CAD has led one company, producing technical instruments, to reduce average turnaround time for printed circuit boards from 35 days in the early 1980s to 5 days in the late 1980s. At the same time, the percentage of boards requiring replotting dropped from over 80% to under 5%. Use of netlist data from the CAD database helped test engineers to eliminate data entry errors and reduce test development time by as much as 15%. In such an environment, competitors that do not stay at the forefront of IT have little hope of staying in the market for more than a few years.

An expert system has been used to analyze electronic testing equipment design, recommending changes to make products easier to manufacture. Over a three year period, it helped cut manufacturing time and failure rates for new equipment by a factor of 4.

Computer manufacturers have taken full advantage of IT. Mainframe development time has been brought down from 3 years to 18 months. Japanese competitors have developed sophisticated new products, increasing their revenues, yet still reducing costs by 50% over 3 years.

A computer manufacturer uses an expert system to analyze customer-specific orders for its products. The system checks the correctness and feasibility of the orders and the availability of parts. It dispatches parts orders to the relevant factories and estimates the expected delivery date. Direct savings result from such factors as reduced stocks and faster billing. Also important is the increased customer satisfaction resulting from on-target delivery and start-up.

Service databases for consumer electronics products are built up with the full details of each product throughout its life—production, assembly, shipping, and behavior in the field. All faults are recorded, as are the actions taken to fix them. This information is made available to all distributors so that when a customer has a problem, they can immediately find out if it is a known problem and then take the corresponding corrective action.

Electronic communication of orders between electronics companies is widespread, and in some cases users of such techniques qualify for extra rebates on list prices for components. Although individual rebates may only amount to a few per cent, they help in holding down overall product costs.

Some electronic component manufacturers claim that as a result of implementing electronic communication throughout the entire order, delivery, and invoice cycle they save nearly $10 per order. On each document, they save 10 to 20 cents. The resulting savings for major electronic companies are measured in tens of millions of dollars.

Manufacturers of professional electronic equipment have been among the first to equip their sales forces with portable computers running a variety of programs ranging from current price lists to time management. In some cases, contact time with the customer, a parameter closely linked to sales activity, has risen by 35%.

Retailers of domestic electrical equipment, used to producing heavy paper catalogs, have moved to other marketing media, such as television and video brochures and Personal Computer disks. Customer databases have been built so that when a customer calls in, a computer-based system scans the customer profile and prompts the sales representative with the best product to propose.

Many retailers now offer their own credit cards. To obtain such a card, customers must fill in a detailed form which provides useful socio-economic information for the store's marketing specialists. When customers use their cards to pay for goods and services, each purchase is recorded, helping to build up a complete customer profile, which is used in defining new sales drives.

3.5 FOOD, DRINK, AND TOBACCO

A distributor of tobacco products installed terminals in stores so that storekeepers could order tobacco goods directly. The storekeepers used the system to order other goods, such as candies and stationery, and the tobacco distributor was able to sell this information onto other distributors.

Now that nearly all goods have bar codes, point-of-sales systems in stores help check-out lines to move faster. Additionally, the information they provide to the storekeeper on goods sold makes re-ordering more accurate and more timely. Storekeepers can use the information to decide where to place goods in the store. Day-to-day information on sales changes the balance of power between the storekeeper and the manufacturer of branded goods, helping the storekeeper to cut the manufacturer's margins. From global statistics the manufacturer used to know better than the storekeeper which goods were selling well. Now things have changed: the stores have the best data. They can use it to promote their own brands at the expense of the branded goods manufacturer.

Aware of the threat, food manufacturers aim to access directly the storekeeper's information, equip their sales forces with lap-top computers to improve their productivity and information retrieval capabilities, and equip their truck drivers with micro-computers so that delivery requirements can be analyzed as soon as possible.

Some companies have opened automated sales kiosks selling limited ranges of goods, such as food, shoes, jeans, and newspapers. These attract new, well-qualified segments of the buying population. Their stock is chosen as a result of forecasting systems that take into account a variety of parameters, such as neighborhood, season, and time-of-the-month.

Traditionally, bakeries have suffered from over-stocking as they tend to cater for worse than average characteristics. From point-of-sale data though, major bakery groups have developed computer models that take account of a range of apparently complex factors in defining delivery requirements. One company feeds weather forecasts into its stock control systems to help identify how customer buying habits are related to short-term expectations of the weather.

As market requirements change, food processors have had to offer a wider range of products in smaller batches and a wider range of packages than in the past. For large companies, the food industry is becoming a high-tech, capital intensive industry. Many of the more important success factors lie outside the classical boundaries of the company. Manufacturers are squeezed between individual customers asking for a wider range of products and supermarket chains with massive purchasing power. They have to produce smaller batches, yet live with reduced margins. R&D costs are rising rapidly to meet the challenge of customer demands for more exotic and healthy products. Global marketing and distribution strategies are needed to win shelf space in stores all over the world for standard products that are fine-tuned to local tastes. The food manufacturing plants are run under computer control with sensors and shop floor data collection terminals providing data that will be used to manage operations efficiently. Planning systems have been implemented and adjusted so that they handle the peaks and troughs of customer demand, rather than maximize the efficiency of the operational process. Production simulation systems allow process engineers to see how incoming orders and expected buying patterns will impact the process. Recipes can be down-loaded in a few seconds through shop floor networks to programmable logic controllers.

Smaller competitors such as small family-run companies in the jams, jellies, and preserves market use IT to help differentiate themselves in the market, hold down costs, or hold their own in partnerships with their bigger brothers.

3.6 MECHANICAL ENGINEERING

Makers of apparently traditional equipment, such as turbines and food machinery have equipped their products with computer-based maintenance systems that continuously monitor operating conditions, inform operating personnel of problems and increasingly try to foresee major problems that could lead to shutdown. This can help reduce unscheduled downtime, maintenance time, and the cost of having service engineers continuously on site.

Air-conditioning system manufacturers provide engineers in office-building companies with computer based systems that can calculate the air-conditioning requirements of buildings of a wide range of shape, size, location, and function. This helps to cement the relationship between manufacturer and customer.

Development of a standardized computer-based design database for a manufacturer of generator equipment led to savings of up to $100,000 a year per part family. It was found possible to support a reduced number of parts and to discount aggressively on high volume orders.

A manufacturer of household power tools recognized the importance of EDI in changing the relationships between a manufacturer, its distributors, suppliers, customers, and competitors. It recognized that by implementing Electronic Data Interchange with its distributors, it would decrease the administrative load on a distributor, leaving time for the distributor to provide better service to the end customer.

A toy manufacturer gave its sales representatives computer terminals to help improve customer service and reduce stocks. Once the sales force had mastered the system, the company was able to cut its order-entry personnel department by 85% and reduce corresponding expenses by 70%.

Many suppliers of parts to major mechanical engineering companies can make on-line access to these companies' production schedules, thus allowing them to have first-hand, up-to-date information on delivery dates. This helps them plan their own production activities, and at the same time, it reduces costly delays for the manufacturers.

A manufacturer of cutting tools with indexable tungsten carbide inserts became aware of the need to find new methods to design the position of the inserts and to generate NC data for insert tooling. In cases where the sculptured-surfaced inserts are arranged in succession along tapered, helical paths, positioning becomes difficult, and the generation of the corresponding NC data a major problem. Introduction of a suitable CAD system and a 5-axis CNC machining center enabled the company to overcome the problem and improve its competitive position.

3.7 METALS, OIL, AND GAS

Oil companies have equipped gas stations with new on-line gas dispensing equipment to ease operation and maintenance, allow faults to be predicted and detected from a centralized maintenance site, and to get sales information back to headquarters faster.

Computer enhanced satellite photographs have proven useful in gathering geological data. Since many of the minerals waiting to be found are in remote areas difficult for people to access, satellites can offer the cheapest form of exploration. Remote scanning techniques and enhanced image processing have led to identification of previously unknown gold reserves. Similar techniques, combined with knowledge-based systems are used by oil companies searching for new oil fields.

As offshore oil extraction moves into ever more inhospitable climates and ever deeper waters, the design constraints on rigs and platforms increase. CAD is used extensively in modeling and testing new designs and ensuring that they are delivered on-time and cost-effectively.

Makers of high-quality steel have found computer-control of their production process to be a major factor in attaining quality targets. In a typical computer-controlled hot strip mill, the time taken to change rollers is reduced from an hour to a few minutes. As the steel comes out of the mill, sensors detect its temperature, and computers control the amount of water sprayed to cool it, helping to tailor the right finish and quality for the particular customer order. A Statistical Process Control system picks up data from the entire operation, processes it, and sends information to the operators. If surface acidity is rising, the operator will be informed and can take action.

Customers can access steel companies' central computers to make new orders, track order progress, and check their financial position. Electronic Data Interchange between the steel manufacturer and the customer cuts days from the order placement to order fulfillment cycle.

3.8 PUBLISHING AND PRINTING: TEXTILES AND CLOTHING

In the publishing and printing sectors, heavily dependent on the dissemination of up-to-the-minute information, IT is increasingly being used to gain competitive advantage. With purely manual systems, it often took two years to get a technical book with a four-year life cycle to market. By the time the book got to market, it was out-of-date and had already lost a large number of potential customers. With IT in use throughout the process, the book can get to market in six months, nearly doubling its useful shelf-life. Being present in the market for 18 months before manually produced competitors, it should more than double sales.

In the past, newspaper and magazine writers would have waded through press cuttings and piles of old newspapers as they searched for the little bit of information that would turn an everyday story into a scoop. Now they use on-line full-text and optical databases to help them get their material to press within the time limits imposed on them by cut-throat competition between press barons.

Some popular and specialist books are now available on compact discs, which can hold up to 250,000 pages of text and graphics. With the compact disc linked to a small computer, readers are given the opportunities to scan for key words, read selected texts, look at still pictures and, in some cases, enjoy the pleasures of animation.

Book publishers communicate information and provide a leisure resource. They are competitors to television. It is not surprising that many of them have identified the global potential of Information Technology and Communications. Some have moved into making television programs. Others have bought into satellites, and beam programs into millions of homes. Spin-offs have sprung up to produce the antenna necessary to receive and decode satellite signals.

In the fashion industry, some middle-market companies can now move clothes off the design board and into shops in less than 2 months instead of 6 months. Orders from shops, with little storage space, are transmitted electronically to computerized warehouse and production facilities. On a daily basis, sales and stocks are examined to define production requirements and to provide basic marketing information. Bar coded labels help reduce the time to get goods from the factory to the customer by cutting down inefficient handling and reducing mistakes, such as loading the wrong truck.

Companies at the high end of the clothing market maintain nearly one hundred parameters about each customer on their databases. Customers can then call up from anywhere in the world and order, say a particular suit in a special material. The basic data is sent to the nearest production site where a computer calculates the most suitable cutting pattern and controls the cutting machine. Customers appreciate the convenience of the process and the speed with which their clothes are delivered.

A multibillion dollar retail and mail order chain suggested to one of its suppliers, a multibillion dollar clothing and textile group, that all orders should be handled by EDI (Electronic Data Interchange). The implementation, estimated to have cost well under $100,000, allowed the retailer to reduce stocks by a few per cent, freeing up several million dollars.

Use of the most effective IT techniques allows a greater variety of products to be produced, yet still be profitable over a shorter life time. Textile manufacturers can accept orders for runs as small as 20% of previous minimum yardage, and they can supply customers with fabrics in half the previous time. Textiles, and woolen goods, such as pullovers, are designed on graphics screens and data fed directly to computerized fabric cutting and knitting machines.

In the automobile industry, CAD is used to speed the design of both car seats and carpet fabrics. Buyers from car manufacturers are invited in to sit at a graphics screen with a designer and develop the required design themselves. Use of high definition screens significantly reduces the number of real samples that have to be produced. Orders can be taken on the spot, and the production process put into gear. Savings on lead time—highly important to car manufacturers—average one to two weeks. Other gains result from improved material utilization, better fitting, and better customer relationships.

Footwear is becoming as much a fashion item as shirts and skirts. Retailers want to be able to order nearer to sell time and to combine shoes and clothes in an overall fashion concept. This puts particular pressure on lead times and distribution channels. The problem is not eased by the very high labor-content in the manufacture of shoes. Computer-based stock control is part of the answer. The product and process development groups of leading manufacturers are also increasingly turning to IT—CAD to help in the design of shoes and CAM to help prepare for their manufacture. In some cases, large buyers such as department stores work with designers on color graphics screens to select the most suitable colors and shapes for coming footwear fashions.

4

SOURCES FOR COMPETITIVE USE OF IT

4.1 TOP MANAGEMENT

This section of the book examines some IT application building blocks that underlie a company's use of IT to support its strategy to gain competitive advantage. In most businesses, it will be necessary to use several of these building blocks. They must be chosen as a function of the competitive strategy, not because it is a good idea to automate a particular process nor that it might be useful to provide some information to a particular manager.

Many of the building blocks correspond to quite common applications of computers. A typical example is CAD. As a system, CAD is CAD, whether the reason for its use is to automate the drafting function or to gain competitive advantage. However, from the strategic point of view there is a great difference. In the first case, the major benefit may only be neater drawings. In the second case, the result may be to win a customer from a competitor, to increase market share, and to secure the company's future. It is the difference between the two concepts that is so important when considering the use of IT for competitive advantage. The fact that a company already has one of the applications mentioned in this part of the book should not imply that it is used for the right reason. For each application the following questions should be asked:

- Why is this application being used?
- What contribution is it making towards competitive advantage?

When developing a description of the various areas in which manufacturing companies use IT to gain competitive advantage, there is a temptation to start with, and concentrate, on the activities that are closely related to the production process and the activities with which the EDP/MIS department has traditionally been associated. In the context of this book however, neither of these approaches would be appropriate. For a book that purports to be aimed at the top management team of a manufacturing company, the most logical starting point is with those activities for which there is a very close link between the use of IT, and top management requirements for competitive advantage. Some of the company-wide IT and communication issues that can provide an advantage, both to individual functions and to the business as a whole, are then examined. These include Office Automation, data management, and integration. They require particular attention from top management as they tend to fall outside the responsibility of individual departments. Since, without a thorough understanding of the customer, the business would not survive for long, the area of Marketing and Sales is then addressed. This leads into a review of the Product Development area and then to that of Production. Potential applications in the support areas of Finance, Administration, and Human Resource are examined. Opportunities for using IT to gain competitive advantage within the Information Technology function are then described. The last area discussed is the external environment of customers, suppliers, distributors, and competitors.

By choosing such an organization for this part of the book, it is not being implied that companies should organize their IT activities along these lines or in this order. Because all companies are different, their organization and prioritization of IT activities will be different. When it comes to the details of how a particular company should select and organize IT applications to support a strategy for competitive advantage, there is no standard answer. Reality is messy. Each company must understand in detail its own competitive situation, and the potential uses of IT. This part of the book describes some of the IT application building blocks and their potential uses. It does not provide an exhaustive list of all IT applications. In view of the number of applications and the objective of the book to be short, concise, and readable, a detailed description of the applications is not given. In many cases, a single line description of an application could be expanded into a chapter. There are many books available that describe applications in such detail. However, the subject of this book is not the details of individual applications, but the overall use of IT to support a strategy for competitive advantage.

Within each area there is considerable potential for the use of IT to gain competitive advantage. However, IT can not, in practice, be boxed into such neatly delimited areas. Communication across the boundaries of the areas is a major issue. There is considerable overlap between use of IT in the different areas. This reflects the real-life overlap of activities between functional areas, which is itself a fertile source of opportunities for gaining competitive advantage through use of Information Technology. There are possibilities to gain competitive advantage not only within each area but also between each area and the others.

For top management, one of the major possibilities to gain competitive advantage through use of IT comes just from recognizing the existence of IT as a potential factor for gaining competitive advantage, including IT in the business strategy, making sure that the rest of the company is aware of the possibilities and enabling them to take advantage of IT wherever possible.

One of the major aims of this part of the book is to make top management aware of the benefits that IT can bring in every function of the company. Because every company is different, the book cannot show managers exactly where they can use IT for competitive advantage. It cannot attempt to go into the details of how each particular application of IT should be implemented. Managers should use this part of the book to provoke thought about the way in which IT can be particularly useful for their own company in supporting the drive for competitive advantage and to improve IT awareness within their company.

There are many ways in which top management can play a key role in developing company benefits from using IT. These include the following:

- Leading the company
- Changing the company culture
- Setting strategy
- Testing assumptions
- Making decisions
- Communicating
- Keeping up-to-date
- Tracking activities

One of the most important is by leading the rest of the company. If the top management team is in favor of IT for competitive advantage and makes use of it on a day-to-day basis, then the rest of the company will soon follow suit. In particular, strategic use of IT by the top management team will put pressure on the IT Department to take a more strategic approach to IT.

Top management can use IT to help change the culture of the company. For example, providing access to external databases can help people realize there is a lot of important information outside the company and that if the company only uses its own, internal information it will see but a small part of the overall picture. Even among top managers, there are many individuals who are so engrossed in the details of current production and sales problems that they ignore what is taking place among customers and competitors.

If a company wants to move towards a product team approach, making extensive use of computers and networks, it can be helpful if the top management team uses the same approach itself. As top managers are often out of the office, they can generally benefit from attempts to use IT to improve their communication and coordination. The resulting everyday use of IT by top managers can help them to develop a better understanding of the IT potential and of some of its associated problems.

IT can help management develop better strategies. It can provide more focused and up-to-date information on markets and products. As shown in Chapter 2, it can help to develop models of the way the company fits into the overall competitive environment and how this information is used and flows within the company. It offers the possibility to look at different scenarios for costs and revenues, investigate the effects of different investment policies on cash flows, and to carry out risk analysis and sensitivity analysis. Models can be set up showing planned progress on major projects. As data comes in on actual progress, the effect of overruns in time and cost can be spotted early and corrective action taken.

Computer-based models provide an ideal environment for management to investigate the effects of new strategies and policies without causing mayhem on the shop floor and among clients. It is much safer to carry out such tests on a model in the computer than in the real-life environment. Many different complex approaches can be investigated in the time it would take to investigate one simple approach without the computer. The effects of acquisitions, different plant locations, and the introduction of new products can all be simulated rapidly. Defense mechanisms against hostile takeover can be modeled and tested. Rapid access to such results will support brainstorming exploratory and intuitive exercises by the top management team.

With IT support, management will be able to make better decisions quicker. The right information will be available, and that will already be a great improvement on the traditional situation. Instead of having to work with ideas and information on several pieces, (or even scraps) of paper, buried deep in out-of-date computer listings, on flip charts, and in a variety of other places, it will be much easier to organize ideas and information. This is clearly helpful because it is often difficult to make any serious approach to a problem until physical access to the information has been achieved. Once the information is available, a wide range of tools is available to carry out the analysis required and to compare results of different alternatives. On the basis of the available information, management can make more accurate forecasts, and carry out analysis to see how reasonable the forecasts are, and which parameters are the most sensitive to change.

IT is an ideal tool to help top management communicate with the rest of the company. There is little point in managers developing first-class strategies, plans, and policies if these are not properly communicated to those who have to implement them. In practice, it is often the problem of communication that prevents successful implementation of strategies. Facilities such as electronic mail can make it much easier for top management to communicate directly with the rest of the company. A letter or memo addressing say, health plans or stock options can be sent from a manager's terminal, and within a few seconds be at the terminal of every employee in the company.

Electronic mail can be used to coordinate meetings among people who are rarely together. It can be used to remind people of due dates for previously agreed actions. It can cut out the need to waste time and money on telephone calls to people who are out of the office. Similarly, communication between management and customers or suppliers can also be greatly improved. Of course, IT only offers the technological foundation for such change. As has been stressed several times, a company has to organize itself properly to gain competitive advantage from such technology. It is not every day that top management will want to send letters to all employees, but IT can help management even in general office activities, such as fixing meetings, circulating memos, and making announcements to selected groups of staff. Communication is a two-way activity, and just as IT can help top management to communicate better with the rest of the company, so can it help the rest of the company communicate better with top management. It is much easier for an employee to send a brief, to-the-point note to a manager on an electronic mail system than to write a formal letter by hand or ask for a meeting through several layers of secretaries.

IT can help top management keep up-to-date on information concerning both the internal environment of the company and the external environment. Management should receive some of this information on a regular basis: some will be required on an ad-hoc basis, with requests for retrieval being accompanied by requests for analysis and, perhaps, more detailed reports.

Information may be needed on competitors, to see who is doing what, how successful they are, and whether this is having any effect on the company's own performance. Requests should be made concerning customer preferences or the way customers perceive the company. The latest information on market place status and trends should be requested. Management may want to keep up-to-date on government activities at both the national and regional levels, to keep aware of the major industrial indicators such as inflation, unemployment, stocks, and customer spending. The deliberations of regulatory authorities and government commissions will need to be followed closely. Newspaper articles and commentaries can be searched for references to key items of interest to the company. IT will offer the possibility for top managers to be in close touch with the external environment. Whether managers seize this opportunity or not remains an open question. Extrapolating the management's requirement to be kept up-to-date, IT can be seen as a vehicle to provide management training. Interactive computer-based training will be available on a range of management subjects, and the manager who lacks experience or needs refreshing in a certain area will be able to access a course through the terminal.

IT will also continue to offer managers the possibility to track activities in the company and maintain operational control. Performance reports will still be required from the various functional areas and exception reports prepared when targets are not met. Again, it will not be the availability of such reports that will lead to competitive advantage, but the ability of a company's management to make use of the available information. The information available to managers can only be what is already available in the company's systems. There will be a great mass of financial, operational, sales and marketing, human resource, shipping, scrap and customer satisfaction data in the system. Managers should be able to do more than read the standard reports. They must be able to ask relevant questions to the system to uncover problems and opportunities. The availability of up-to-date company status, functional performance, and individual project progress will provide them the foundation for improved control. They will be able to go further to unearth the reasons behind specific problems and opportunities, such as missed project milestones or spectacularly improved sales results in a particular region. It is probably at this level that they will start gaining significant benefits from the system.

The ready availability of up-to-date information can greatly increase the productivity of management meetings. Only too often, in these meetings, top managers ask questions for which their subordinates do not have the answer on hand: this reduces the effectiveness of the meeting. By the time the answer has been dug out of thick piles of listings a few days later, the top manager is probably at another site, and an important decision has been delayed by a week or more. If the information could be instantly available then time would not be wasted and prompt action could be taken.

Attempts have been made to develop integrated systems that would offer top managers the various types of assistance previously mentioned. These systems are generally known as Executive Information Systems (EIS). Such a system must be able to access whatever information is required by an executive both inside and outside the company. As a result it is usually built around desktop computers that are part of a company-wide network offering the possibility of communicating with external databases. EIS will typically have a high quality, easy to use, graphics interface so that users do not have to spend months learning unintelligible languages. Graphics output is a must for the fast communication and assimilation of information. An EIS should be able to display an "operating indicators" screen so that executives can see at a glance the current situation of the company.

Some companies are already using such systems. Although many managers initially find them difficult to use, they persevere, knowing that they are learning how to use such systems, becoming aware of how such systems can be of use to the company, and setting the foundations for using them to gain competitive advantage.

4.2 COMPANY-WIDE INFRASTRUCTURE

Many opportunities for gaining competitive advantage through use of IT are found in individual functional areas, at the interface between functional areas, and between groups of functions. It should not be forgotten that there will also be opportunities at the level of the whole company. It will not be possible to take full advantage of these unless a certain degree of standardization is applied. Without this the efficiency of both systems and people will be reduced. For the same reasons that companies try to standardize their activities that are not related to IT, they should also try to standardize their IT-related activities. Without very good reason a company would not equip itself with differing and nonstandard machines, tools or accounting systems. The same is true of IT equipment and systems.

Some companies will choose to develop their own standards and adjust systems that they purchase to meet these standards. Some standardize by trying to purchase all IT equipment from one vendor. (Although this may actually lead to a nonstandardized environment if the vendor is unable or unwilling to completely standardize its own products.) Other companies will prefer the route of international standards. For the IT environment of the 1990s, this appears to be the most suitable situation. This runs contrary to the strategy of some IT suppliers, who claim that their systems are vastly superior to those of their rivals. From the point of view of most companies using IT, it is unlikely that a particular system will bring long-term competitive advantage. Rather this will come from organizing the company to meet market requirements. The effort that is needed to do this is such that a company will not want to waste energy on understanding how to overcome the technical difficulties of nonstandard systems.

At the basis of the company's IT strategy should lie detailed company models describing the business environment and the flow and use of information. These models should exist for all the company, not just individual functions, and should include details on the external environment. Without such models, much will be lost: there will not be a unified view of the overall company, and it will be difficult to identify the best way to use, store and communicate information. People will continue to carry out duplicate and unnecessary tasks and there will be a tendency to design systems that fit within boundaries of individual functions even though these may be completely inappropriate. Interfaces between functional areas will waste time and effort and continuously produce errors. Top management should make sure that models, even if they are not particularly detailed, are developed as early as possible for the whole company.

An infrastructure, a technical foundation, should be established to support the use of IT across the company. The technological IT components will continue to change rapidly, yet the company will not want to have to continually change its whole approach to IT. To avoid this, a technological infrastructure, based on basic concepts, rather than on specific technologies, must be defined. It should be flexible enough to handle the many changes, some quite drastic, that will occur over the next 5 to 10 years. The four basic elements of the infrastructure correspond to the four basic activities of processing, moving, storing, and handling information. It would be wasteful to develop different information infrastructures for each individual function of the company. For example, there is no reason why people in the company should not handle information through the same user interface. Yet, today, most functional computer systems have completely different interfaces, and much time is wasted in training people just to get to know how to approach a system. Since some people may not be able to work with several interfaces, and others will just refuse to learn another user interface, the advantages of a company-wide information infrastructure are numerous.

In the same way that a company-wide infrastructure will be much more beneficial than a set of disjointed functional infrastructures, the approach to the information architecture and the systems architecture should also be company-wide. A major requirement for the infrastructure is that the IT equipment and systems in different parts of the company can be connected together in a secure, efficient, and manageable way. Otherwise information will not flow between the various functional systems. Sales forecasts will not be translated immediately into probable purchase orders, therefore time will be lost, errors introduced, and costs will rise.

In the future, companies will find themselves handling enormous volumes of information. Large corporations will need to manage volumes approaching 10^{16} bytes. It is very unlikely that any company is going to have all of this data stored in one single location. The environment will almost certainly be one of data distributed on a wide variety of media, in a wide variety of locations on a wide variety of systems running on a wide range of computers from a wide range of vendors. Without a company-wide data management strategy and the corresponding tools, chaos will ensue. Historically, the computer has offered two main advantages; the ability to process data much quicker than humans, and the ability to store and access much larger volumes of data than humans. In some functional areas, for example CAD, the ability to recall data rapidly (e.g., reusing an existing part with a minor modification to meet a new client requirement, without requiring redesign from zero) is cited as a major justification for use of the system. Yet with the volumes of data that exist in many companies, if the data is not managed properly, it becomes difficult and time-consuming to find and access the specific set of data required.

On a company-wide basis, it will be necessary to define who has the right to access data, who has the right to modify it, and who has the right to archive and manage it. Each item of data will have to be classified as company-wide or function-specific and its ownership clarified.

The user interface is another important part of the company-wide infrastructure. If it is possible, the same user interface should be used throughout the company. Everybody should be able to gain entry to the system through the same interface at each workstation (illegal access will be prevented by the system's internal security controls). The user interface should be updated as new developments in AI and voice processing open up new possibilities that the company will want to exploit. However, obvious benefits arise if the same user interface is used throughout the company, and everybody in the company knows how to use it.

An apparent example of company-wide use of IT is Office Automation (OA). This is probably not a typical example though because many of the "solutions" available can be used as stand-alone applications that are not closely linked to previously installed systems. In future OA solutions, it will be necessary to integrate in-place systems, and this will be more difficult.

Office automation systems provide an example of the type of IT solution that will appear in future. They address the infrastructure components previously mentioned and are based on widespread communication of information, distributed use, storage and processing of data, and a common user interface. Where they do not use industry standards, they are of course creating another discrete component of the overall company IT jigsaw, which one day will have to be put together.

These systems offer a range of applications that can include word processing, electronic mail, electronic filing, project management, data communication, and even voice communication. These applications correspond to activities carried out by a wide range of people in the company. An engineer uses word processing, as does a top manager and a secretary. All three may make use of different features, such as checking the spelling of words. Similarly project management can be used to organize an individual's activities, the activities of a small functional project team, or a major project spanning several functions. Graphics capabilities and spread sheets available with such systems will also find many users. The sales manager uses graphics to show top management the sales results in different areas. This is much more effective than producing long lists of unreadable numbers. Sales teams use graphics to include diagrams in proposals. Production managers will be able to track their production and scrap figures on a daily basis rather than waiting several weeks to receive the results back from headquarters. They will be able to compare their actual performance with planned performance, carry out trend analysis, look at moving annual totals, and try to understand how to improve performance. Working on data that is several weeks out of date may be totally useless, as in the meantime other parameters may have changed. Working on fresh data provides a closer link between actions and results.

Spread-sheets will clearly be useful to financial analysts, but they can also help product development teams to better understand the financial issues associated with their projects, and allow top managers to look at the effect of different scenarios for company strategy. Office automation systems often offer the possibility to set up "electronic diaries" for those who need them. These can be helpful for individuals in planning their own time, and also in arranging meetings with others. People in widely spaced plants can use electronic mail to keep in close contact with other plants and with headquarters.

Most people make regular use of information of a particular type and want to have instant access to such information. However every now and again they require other information. It may be useful to maintain a set of lists that will help them find what they need. Lists of magazines and journals received by the company can be useful. So can lists of employees previous employers in helping to find entry points to other companies. Also a list of company and competitor products together with a list of contact names in customer companies can help remind people what they are competing against and may stimulate them to gather information on competitors.

The potential benefit to be gained from use of IT on a company-wide basis is tremendous. However unless top management is involved in directing its use, it is quite possible that no real benefit will occur. It is only to easy to let the IT specialists set up their personal "diaries," confidential databases, individual letterheads, lists of private contacts, and so on, on their "own" computers. Before long each of these experts will be buying incompatible equipment from different vendors in an effort to have the "best" equipment. This type of behavior and activity is not likely to lead to competitive advantage. However it will occur unless top management lays down the ground-rules.

As with all other activities, top management must set the goals for company-wide use of IT. The goal of office automation is *not* to type twice as many letters a day, to show pie charts of sales penetration in different industries, or to have automatic telephone call-up from a personal computer. The goal of office automation must be business-related. Examples include increased sales, increased quality, lower overhead cost, and increased time before the customer. Unless top management makes it clear that this is the case and that the systems will not be replaced in a few years' time, but are there for the long-term, then they will not lead to long-term competitive advantage.

Before introducing "company-wide" applications, such as Office Automation, top management must examine very closely the cost-justification of the application. These questions should be asked: Where will cost savings be made? What parameters will be used to measure them? Where will increased sales occur? How can they be identified? How can increased sales be directly linked to the introduction of a new "company-wide" application?

To summarize, the following are some company wide issues which affect the drive towards competitive use of IT:

- Industry standards
- Company and information models
- Infrastructure and architecture
- Integration
- Communications
- Data management
- User interface
- Office automation

4.3 MARKETING AND SALES

A major requirement for successful Marketing and Sales is information. Information is required about the company itself, including its ability to develop and modify products, and on product price and promotion. It is also required about customers, markets, and competitors. Much of the internal information requirement overlaps with that of the other departments in the company. Marketing must know how long it will take to get a proposed new product to market: Engineering and Manufacturing will heavily influence the answer. Each department will use historical information it has collected concerning previous similar products to estimate future time cycles. Similarly, much of the other information—sales, costs, stocks, and so on will be shared with other departments.

Information on the overall market, economic conditions, and government and regulatory policy is needed. Geographic Information Systems linking computerized mapping, land registration, and socio-economic data on customers can be used to pinpoint the best locations for new sales and service outlets. Linked into credit card sales they help to build up detailed pictures of zonal customer profiles and patterns of spending. For each customer, comprehensive details should be readily available. Potential customers can be mailed at their home address. Similarly, point-of-sales systems are a useful source of information for marketing staff. Instead of waiting several weeks to get reports on product performance, up-to-date information is available, helping to build up an accurate picture of customer behavior. Other information on customers may come through market surveys or from distributors. Competitive intelligence is necessary to understand competitor's strategies and behavior. This can help in preparing a defence against a competitor. Alternatively, it may be used as a source of comparison to identify areas where the company is currently weak.

Few companies would disagree that the type of information previously described is useful, yet few companies are able to handle this type of information effectively. All too often, one hears that there is too much wrong information available and not enough right information. This is partly due to the unwillingness of most companies to invest the necessary time and money in understanding what type of information they really require. Again this comes back to the need for a company-wide information model that has been mentioned several times before in this book. Many companies find that in several departments, attempts have been made to build up models describing the local use and flow of information. Unfortunately, these are rarely brought together with the result that most of them are incomplete and erroneous. Marketing's need for information shows how important it is for the company-wide information model to extend beyond the boundaries of the company and include information on the market, customers, and competitors.

Once a company has managed to gain control of its marketing information, it then becomes possible to start using it to improve strategic decisions. Market research data can be analyzed faster and in more depth. Approaches such as time series forecasting and regression analysis can be applied to predicting sales, matrices can be built up showing how features of the company's products compare to features of competitor's products, price and sales data can help understand competitor's strategies and customer's requirements, and screening and analysis of the business case for new products can be carried out in more detail. Also models can be developed so that new product strategies can be tested under different market conditions and with different percentage market growth rates. Some companies build up extensive customer profiles in their data bases to be able to test how strategies would be received by customers, and how they would affect market share.

Computer systems have not yet taken over from the human specialists in Marketing. However they are increasingly being used to support decision-making. Often this is done by sifting through the great volumes of market data available from scanner output. Without the help of the computer, it is virtually impossible to analyze all the information now available. Marketing decision-support systems are used to fix prices, establish where distributors would be most effective, understand the effect of different forms of advertising, and foresee problems arising from competitor strategies. On a more day-to-day basis, they can help plan sales campaigns, show which sales strategy is proving most effective, sharpen promotions, identify which customers should be approached next, and suggest the most effective way in which individual customers should be approached.

The potential of IT may lead to new strategies in itself. One of the best examples of this is when a manufacturer and a retailer link their systems. Another example is the installation of a terminal in a customer's purchasing department. This will make it easier for the customer to buy, and hopefully lead to more sales. It should greatly increase the speed of communication between the two companies. The customer will be able to access up-to-date product and delivery information and the company will receive information on the customer's buying intentions, and over time build up a good picture of buying behavior. As well as developing a closer link between the two companies, introducing such a terminal can also raise a competitive barrier. Competitors without a network will find it difficult to reply. They will have neither the technological know-how nor the technology readily available. Even competitors who do have these available may find that companies are unwilling to install several terminals since there is a large hidden cost in training involved.

The sales manager can make use of IT to optimize the size of the sales force as a function of the number and location of potential customers and the necessary frequency of contact. Once this has been done, balanced territories can be allocated to individual sales representatives. As data is fed back into the system by the sales force, the sales manager can review the efficiency of representatives and of relationships with customers as a function of actual sales. Analysis of individual customers can lead to redistribution of sales resources and efforts. The results from foreign countries can be analyzed faster, and the resulting actions taken quicker. IT can help a company have a global strategy, but act locally. Potential and existing clients, even if on the other side of the world, are only a few seconds away on a communications network.

Telemarketing is an area that is heavily reliant on IT and communications. In the absence of face-to-face contact with the potential customer, the salesperson must be armed with information, and prompted by the system on the products to be proposed. Results of the conversation should be fed back to the system so that they can be used in the next contact with the customer and in building up the overall market-profile. Similarly, telephone interviews can be assisted by IT. The questions can be set up on the interviewer's screen and answers entered immediately. As a function of the answer, the system can propose which question should be asked next.

Sales quotations can be significantly improved by assistance from the computer: they can be produced more accurately and quicker.

More reliable cost quotations can be provided for complex products. They can form part of an attractive proposal produced by a word-processing system in minutes rather than in days. When the system is extended to include a CAD system, a very powerful tool is available. Complex customer-specific proposals can be produced in a fraction of the time previously required, yet contain all the text, numbers, and graphics that the customer expects to see.

Computer-aided quotation systems are often included in the set of tools that are now often offered to sales forces equipped with portable computers. A sales representative can take such a computer into a meeting with a customer and have available a mass of information that previously was only available back at the head office. The system can have basic information on products, prices, promotions, discounts, customer history, and so on. The sales representative has information available on the full product and service offering but can focus on those areas where past history proves that the customer is now likely to have a need. An up-to-date quotation can be made on the spot. During the meeting, the sales representatives can enter information into the system about customer requirements and customer views on the company's products and those of competitors. Later this information can be transferred to the company's Marketing and Sales information system where it will be used in developing new strategies and tactics. The portable computers used by sales forces contain a range of other programs. These can help to schedule day-to-day sales visits, helping to improve the amount of time the sales representative spends in front of customers. They can also bring new products to the attention of a sales representative at the most relevant moment. This may be just before a meeting with a potential customer and can motivate the sales representative to get acquainted with the new product and to try to sell it. It overcomes the age-old problem of new product information being received (and filed and forgotten) when released by the product development team rather than when needed by the sales force. The system can also help the sales representative track prospects, maintain a history of proposals and follow-up actions, and evaluate the risks associated with each proposal.

Orders, whether they came from salespeople in the field, from distributors, or directly from the customer, must be properly managed. Sales order processing systems handle this activity. Although their concept is easy to understand, in companies with large order volumes, they can be difficult to run efficiently; as the number of orders, the numbers of different products and parts, and the number of sources of parts increase significantly, their operation becomes very complex. As business is handled through more and more channels and across different sales area boundaries, errors become more frequent and, for example, major customers may mistakenly be given the wrong discount.

Sales forecasting systems help manufacturers to foresee the demand for materials and parts. The more complete and up-to-date the information on actual and expected sales, the more accurately projections can be made on future sales, and hence the need for manufacturing and distribution resources. Although it is possible to develop forecasts manually in simple situations, most companies find themselves in a more complex environment.

In this case they become involved with techniques, such as time series forecasting and regression analysis, which benefit greatly from computer support. As the analyses become more complex with additional factors being added, the ability of the computer to carry out large numbers of calculations and produce graphs and charts automatically is greatly appreciated.

Sales analysis systems are at the forefront of understanding the effectiveness of individual products, sales representatives, promotions, strategies, and so on. They may show market share falling, an increasing fraction of sales made through indirect channels, or a steep rise in orders from a particular customer. Such analysis is essential for understanding the difference between planned and actual performance prior to developing new sales strategies and tactics.

IT can play an important part in sales force training. In those companies where sales forces are equipped with portable computers, some training may be carried out on these. Back in the office, interactive video training may be used. IT can be used in product training, in sales skills training and in improving understanding of markets and customers. Basic training on products can be in the form of question-and-answer sessions on the screen. Skills training may use computer-based models to simulate sales situations and allow the sales representative to investigate the effect of different approaches.

The following are major IT issues in marketing and sales:

- The basic Marketing and Sales information
- The Marketing and Sales information model
- Marketing strategy
- Marketing Decision Support systems
- Sales strategy
- Sales management
- Telemarketing
- Sales quotation
- Sales force automation
- Sales order processing
- Sales forecasting
- Sales analysis
- Sales force training

4.4 PRODUCT DEVELOPMENT

More than thirty years after computers were first used in engineering, it could be thought that great productivity gains would have been achieved in this area and that there would not be much more progress to be made. Yet the situation is completely the opposite. Very few companies have made significant productivity gains through the use of computers in product development, and there is still a lot more progress to be made. Some companies have of course benefited greatly from the use of computers in product development: for example; in aerodynamic analysis of new aircraft; in the design of sculptured surface parts, such as helicopter rotor blades, car bodies, and turbine blades. Electronic circuits of today's complexity just could not be produced without computer support. Yet in the majority of companies that do not make high-tech products it is often difficult to find an example of the use of a computer in product development that has resulted in a significant improvement in productivity measurable at the level of the bottom line. Such statements will appear heretical to many suppliers of computer and engineering application packages. Yet paradoxically such companies are partly responsible. They have sold innovation-led products without considering their customers overall needs. Before long manufacturing companies may find themselves in the situation that continued use of applications such as CAD will lead to significant drops in quality and productivity.

At the heart of the problem lie two major issues—the difficulty of managing engineering data effectively and the need to modify the organization of engineering activities to take advantage of the introduction of IT. Too often, companies have installed individual applications that promised to offer local productivity gains without causing major problems. CAD is such an application. People were astounded at the speed with which a drawing appeared on a screen or could be produced with a pen or electrostatic plotter. They imagined that their design process could be speeded up by a similar factor. In practice it cannot, because only a small fraction of the design process involves producing paper-drawings, which in any case may be superfluous. In many large companies there are more than thirty different applications running in the product development function. Some of the systems run on mainframe computers, some on super-minicomputers, some on engineering workstations, some on Personal Computers. People transfer data from one application to another by hand—wasting time and introducing errors. It is not difficult to understand why, at the end of the process, there may be no real productivity gain.

Few companies have been prepared to stop, review the situation, decide on the best long-term strategy, describe the policies, define the architectures, cut out unsuitable systems, buy new hardware and software, and retrain people. Even companies that have taken such measures find themselves confronted with a major data management problem.

There are vast volumes (such as 10^{12} to 10^{15} bytes) of engineering data in large companies. Data are on different media—paper, disks, cassettes, and tapes. Data are generated in different formats.

The applications will go through several versions. They run on different releases of operating systems on a variety of computers. The data are in the form of graphics, text, or numeric. The data will "belong" to different groups, perhaps different departments.

To add to the complexity of the pure data management problem are other problems that arise from trying to meet customer requirements for short development cycles, short product lives, and customized-products. These requirements lead to greater numbers of product configurations and of product and part versions. In turn, these increase the complexity of managing engineering data.

In practice, unless care is taken it becomes very difficult to know for example exactly which spare part needs to be urgently sent by overnight mail to Customer A13 whose X225 product, serial number C165269C, and purchased on 13 May 1981 requires urgent repair. What CAD system did we use then? Did we archive it? Will the data be usable under today's release? Did C165269C have BD-16-8-48 or BD-1-2-74? Didn't we change about that time? Where did we put the field test results? Do we still have that computer? Is the master data on tape or on paper? . . .

The problem is getting worse and worse as more and more engineering personnel use computer-based systems that produce hundreds of megabytes a day, yet offer few management controls against deliberate or accidental mistakes. The answer to the problem lies with engineering data management systems. These are of utmost importance for engineering activities in the 1990s.

Having all the engineering information under computer control simplifies the control of engineering changes and allows project managers to more easily monitor the resources and time scales of the design phase.

As the computing environment becomes more and more distributed, and company management attempts to gain control of engineering data, communications networks between engineering computers, workstations, and terminals will become much more common. This will be particularly true for companies with engineering operations in several countries. Few engineering applications or activities are completely unrelated to all others, and whether they need to transfer data or simply project status information to and from other applications, the fastest and most accurate way is over a network. The only exceptions to this general rule are those related to security.

The problem of data exchange between systems using different data structures will continue. This problem is clearly seen in the world of CAD, where each CAD system generally has its own data structure, and one system cannot read the data coming directly from another system. As a result attempts have been made to define international standards for data exchange (such as IGES—Initial Graphics Exchange Specification). In principle, data from CAD system A is translated from its internal data format into IGES format which can then be read by system B and translated into its own internal data format. Unfortunately, as with all translations, the result at best can only be 100% accurate. In all other cases, it is less than 100% with some information misinterpreted or lost, and an associated significant time lag. Companies in the 1990s will be under pressure to reduce development cycles. They will not want to work with data exchange mechanisms that cause them to lose time and information. The 1990s may well see a swing away from international standards for data exchange that must be able to handle many different types of data entity, to the increased use of specific translators that only apply to a limited number of entities.

The three areas previously described: engineering data management, engineering networks, and engineering data exchange, are all related to the company's business and information models. It is not possible to manage data effectively unless it is known where, how, and why data are being used, or why and how it is communicated to other activities. Developing the information model corresponding to use of information in the product development function provides a company with the basic knowledge necessary to implement an effective data management and communication system.

The engineering information model must not be developed in isolation from the rest of the company. Engineering activities are closely linked to all other activities of the company, and engineering information will be used throughout the company. Clearly this is true of Manufacturing, which will make use of Numerically Controlled machine programs developed in engineering. It is also true of Manufacturing Planning which requires Bill of Materials information. Sales representatives may want to use engineering data to make a drawing of a product that will be sent to a potential customer. The information model developed for Engineering should take account of all the interfaces with activities in other functions in the company and where appropriate outside the company.

Alongside purely technological issues of using computers in product development activities are associated organizational issues. Two very important, and related issues, are the procedures governing use of systems and the training of the users. It should be obvious that people who are not properly trained will not be as productive as they would be if they were properly trained. Unless formal procedures are developed showing people how to use systems, then people will use them as they see best. There are times when it is sensible to give people free rein to let ambition and imagination lead them to successful, innovative products. However most of the time, work on engineering systems occurs in a relatively structured environment, and the introduction of procedures does not inhibit innovation but prevents anarchy—the result is a saving of time and energy.

To understand the justification for training, some wider issues, looking further than the cost of training must be addressed, and the aims of the engineering function must be understood. It may become clear that a six-month design cycle is required and an eight-month cycle would lead to lost sales of $5 million. Providing suitable training may cost a few tens of thousands of dollars—say even $100,000—yet this may be all that is needed to attain a six-month design cycle.

Far too many engineering companies spend millions of dollars on technology for product development systems, and try to make up for it by saving the few tens of thousands of dollars that would be required to train people properly and implement suitable working procedures. In so doing, they lose the company several more millions of dollars in lost sales—but of course they see that as a Sales problem, not an Engineering problem.

CADCAM and CAE (Computer-Aided Engineering) are two of the most important systems in the product development environment. Although CADCAM is most often associated with mechanical engineering, and CAE with electronic engineering, conceptually they are very similar. They both refer to the application of computers to the design engineering and manufacturing engineering process of a product, and therefore, refer to the total engineering function of the company. Their aim is to increase the quality, flow, and use of engineering information throughout the activities of defining what the product is to be, and the way that it is to be produced. This translates more concretely into reduced cycle times, reduced cost, and increased quality. These are the types of competitive advantage that most companies are looking for.

The use of CADCAM in mechanical engineering can be taken as an example to illustrate some of the important features of the use of computer technology. First, there is the very important aim of modeling the part in the computer and of reusing this information at later stages of the engineering process. Geometry modeling is the process of building a model (in the computer) that contains all the necessary information on the part's geometry. The model should be unique (so that the part will not be mistaken for another) and complete (it should contain all the geometry information required in later activities). Product modeling covers the process of building up a computer-based model containing all the necessary information on the part or product. This information includes other attributes apart from geometry, such as color and material.

Another characteristic of the use of computers in product development is the use of an interactive graphic screen or workstation. There is also the concept of a set of application programs, which carries out specialized tasks in particular areas. These include finite-element analysis and NC programming. Engineering drawings can be built up on the graphics screen and, if necessary, modified. Once the drawing is satisfactory it can be automatically drawn on paper by a plotter. Another application that benefits from the use of CADCAM is simulation. A designer may investigate on the screen, and from a variety of viewpoints, how a mechanism will move and whether it will interfere with another part. As well as aiding the programming of NC machine tools, such as milling machines, CADCAM can be used to program robots and quality control machines. Tool loads, feeds and speeds can be optimized. In the manufacturing engineering area, it can also be used to aid the design of tools and the preparation of process plans. Programs can, for example determine the best shape for dies for forged parts.

CADCAM is a very useful computer-based application making use of interactive graphics techniques. It can be used for many applications throughout design engineering and manufacturing engineering. However a CADCAM system is only a tool. It does not automatically provide competitive advantage, nor does it automatically design, analyze, or manufacture. These tasks must still be carried out by people (managers and designers), programs (finite-element analysis), and machines.

One of the most important features of CADCAM is the potential it offers for reuse of part data and the associated need for a computer based model of the part that can be used in several application areas. In the past, with manual techniques, information on part data was transmitted from person to person on drawings. Manually produced drawings of typical mechanical parts do not always exactly reflect what the part is—they tend to be incomplete, ambiguous, and incorrect. CADCAM systems do not currently have the intelligence to decide what the person who produced the drawing was really trying to describe. Even if the drawing does give a meaningful description, it may not contain the information required by another application area. Electronic transmission of a complete CADCAM model of a part is a surer way of transmitting information than a drawing on paper.

CADCAM can enable more designs and products to be produced within a given time frame. It can be used in this way to gain competitive advantage. It can be used for a significant amount of design work in projects and, hence, shorten the engineering cycle. CADCAM can also lead to an increase in the reuse of existing designs, thus reducing the cost of the design process. Used in conjunction with Group Technology, it can lead to a reduction in the number of parts that are needed to produce a wide variety of customized products. This reduction will, in turn, reduce process planning costs and lead to a reduction in manufacturing engineering and inventory costs. All these gains can be directed towards attaining competitive advantage.

Use of CADCAM can result in a variety of benefits. Many of the analysis programs involve carrying out a large number of calculations that just would not be feasible by hand. The use of 3-D design and display of a part on a graphics screen gives the product developer the possibility to identify errors in the early design stages rather than in manufacturing, when major costs have already been incurred. Reuse of the same model of the part by different application programs reduces the time previously wasted in reproducing and modifying drawings. All these gains can help in achieving the overall goal—the attainment of competitive advantage.

As more experience is gained with the use of IT in the product development function, the focus of interest changes.

In the 1970s, the potential view for using computers in product development was as follows:

- Computer Aided Design
- Numerical Control programming
- Finite Element Analysis
- Computer Aided Process Planning
- Simulation
- Computer Aided Technical Publications
- Programmable Logic Controller programming
- Robot programming
- Automated Testing Equipment programming
- Machinability Data Systems
- Parts List Generation
- Computerized Piping, Nesting, Die Selection

The view of IT was very much a set of individual computer systems, each of which would provide a major productivity gain. In practice, in most cases, such systems have not, for various reasons, provided the expected gains. Invariably this has been because the organizational issues have been ignored. Questions such as—Why are we using this system? and How should we use this system? have been bypassed in the rush to use the system.

In the 1990s, the outlook for using computers in product development is foreseen as follows:

- Engineering data management
- Product modelling
- Engineering data exchange
- Information models
- Integration
- Engineering system procedures
- Training

It has been recognized that individual stand-alone systems are unlikely to provide major productivity gains at the company level. Instead, the first question that is asked is how can the company use IT to gain competitive advantage? Only then can it be asked if it is possible that this system can help to gain competitive advantage? If yes, then the company must understand how the system can be used within the context of the entire entreprise for this end. Questions are asked about information flow and the various activities that take place along the product path from design to manufacturing. Only if a system can be seen to fit into the overall picture, and can be integrated with other systems and activities, can it be considered.

4.5 PRODUCTION AND DISTRIBUTION

There is probably no function in the typical manufacturing company that has been so open to automation and computerization as the production function. A wide range of applications has been developed to address the activities of manufacturing planning, manufacturing control, physical manufacturing, and distribution.

The approach has been piecemeal. On the one hand, systems have often been developed by small specialist companies aiming to provide maximum productivity from use of their systems in narrow, very well defined areas. On the other hand, systems have often been selected and implemented by small groups of people in the company requiring justifiable solutions to well defined problems. The result has been the implementation of a large number of "point" (or "island") solutions that improve local productivity but often lead to an overall loss of productivity when information and materials are transferred from one island to another.

In the production function, manufacturing companies have taken the traditional approach to using computers. They have implemented systems to automate existing operations and to provide improved management control. This approach will not necessarily lead to competitive advantage. High quality products and short manufacturing cycles are the objectives that top management must bear in mind when using computers in the production function as a source of competitive advantage.

The islands of automation in production and distribution include:

- Material Requirements Planning (MRP)
- Manufacturing Resource Planning (MRP2)
- Bill of Materials Processor (BOMP)
- Stock Control
- Purchasing
- Manufacturing Simulation

- Shop Routing
- Short-term Scheduling
- Shop Floor Data Collection
- Receiving
- Bar-code Reading
- Machine Monitoring
- Personnel Monitoring
- Material Handling Systems
- Numerically Controlled Machine Tools
- Computer Numerical Control (CNC)
- Direct Numerical Control (DNC)
- Flexible Machining System (FMS)
- Automated Assembly
- Automated Inspection
- Robotics
- Programmable Logic Controllers (PLC)
- Automated Warehousing
- Process Control
- Automated Guided Vehicles (AGV)
- Distribution Resource Planning (DRP)
- Computer Aided Plant Layout
- Work In Progress (WIP) Management
- Automated Storage/Automated Retrieval (AS/RS)
- Coordinate Measuring Machines (CMM)
- Statistical Process Control (SPC)
- Computer Aided Quality Assurance (CAQA)

Rather than investigating these systems, management should be looking at the following production issues:

- Manufacturing Strategy
- JIT, TQM, and SMED
- MAP
- Information model
- Integration
- Factory management systems
- Simulation
- Training

Without a well-defined business strategy containing a detailed manufacturing strategy, any investment in computers in the production function is likely to be ineffective when considered from the point of view of competitive advantage.

Just In Time (JIT), TQM (Total Quality Managment), and SMED (Single Minute Exchange of Dies) are all examples of successful approaches that can be taken in Production and do not require the use of computers. They are examples of simplification before automation.

The acronym JIT covers a wide range of different approaches. JIT can imply a company-wide philosophy of waste reduction. In a more limited sense, it may just be taken to mean a specific approach, based on actual usage, towards the transportation and processing of material and parts. The material should be transported "just-in-time" to the machine where it is required for processing. It is the need at the machine that drives the need for transportation. This is a "pull" approach, rather than the "push" approach of Manufacturing Resource Planning (MRP). MRP is a forward planning system which prepares material for the manufacturing process on the basis of forecasted demand. In the extreme case, an MRP system will build up stocks in front of each processing center to make sure that the centers are always busy. As these stocks have to be managed, tracked, identified, and stored the whole process becomes very complex. On the other hand, in the ideal situation, there is zero stock in a JIT situation as material is either being processed or being transported to the next processing operation. (This is not to say that JIT is superior to MRP. In most manufacturing companies both will be needed, MRP as the forward planning system and JIT as an operational system. Some companies will have no need for MRP and others little need for JIT.)

TQM, like JIT, is an acronym with many meanings. It can be a philosophy for running an entire company. On the other hand it can be an approach for reducing defects and scrap. There are good arguments for removing all inefficiencies from the production process before automating it. Why buy a large computer system to help manage scrap? If there is no scrap, then the money that would have been invested in the system can be invested more fruitfully elsewhere.

As production runs become shorter, change-over time becomes more and more important. Change-over times of seven to ten hours become unacceptable in environments where production batches only last a few shifts. They cause too much valuable production time to be lost. Approaches like SMED, which aim to significantly reduce change-over times, can sometimes improve productivity much more than the introduction of a new computer system.

The strategy must meet the business objectives of increased productivity and quality for today's environment and increased flexibility and adaptability for the future environment. It must address the siting of production and distribution facilities along with plant and warehouse layout. It must also show which processes will be used, how they will be used, and what equipment will be needed.

The communication of information throughout the manufacturing facility is a major requirement. Information has to be transferred to Manufacturing from other departments, such as the Product Development Department. Information also has to be transferred between different activities in the manufacturing facility. The initial response of most nonstandard, proprietary IT systems is to allow communication between all of the vendor's equipment. However most manufacturing companies have a range of equipment from many vendors and want to transfer information between equipment from different vendors. Manufacturing Automation Protocol (MAP) contains a set of rules showing how machines from different vendors should communicate together.

A communication protocol is just the beginning of the information communication solution that a manufacturing company needs. A communications network is needed, and at the facility level this will be a LAN (Local Area Network). Before implementing such a network, an information model should be built up showing what information is being used and where and how and why it is being used. The model should be part of the company's overall information model, and show clearly the information interfaces between the production function and other functions. The communications network should reflect not only the current use of information but also the expected use in the future.

Integration is just as important an issue in the production function as it is in other functions and between functions. The integration of information is a key issue. Of equal importance is the integration of material. Just as information should flow freely between different activities, so should material: again this implies the removal of stocks.

Production management needs information, such as Bills of Materials and methods and routings from the Product Development function. Information is needed on product forecasts, finished-goods inventory, WIP, capacity, order backlog, and so on.

Many of the manufacturing planning functions have been assisted by computers for several years. In discrete manufacturing companies, the use of computers on the shop floor has occurred more recently. Examples include shop floor data collection and management, tool management, real-time control, short-term scheduling, fault diagnostics, and maintenance.

Automated warehousing systems offer functionality in areas such as storage and retrieval, warehouse enquiry and stock amendment, and warehouse status and stock reports.

Distribution management systems help to site warehouses as effectively as possible, balancing the need to be close to customers with financial constraints. Transportation is scheduled as a function of customer requirements and locations, and vehicle availability.

The introduction of computers into the production and distribution function creates a need for training of production and distribution personnel. In some cases this may just imply the need to be able to use a terminal to carry out the tasks that were previously carried out by hand. In such a case the major difficulty encountered in training maybe in getting someone unfamiliar with electronics and computers to adjust to working on the screen. In other cases, though, where the introduction of the computer is part of a major change in the way of working, the training need becomes significantly more complex and time-consuming.

IT can affect relationships with key customers and suppliers. The focus of each plant must be examined. It may be useful to focus each facility on a particular product line. Alternatively, within a facility it may be useful to set up "subfacilities" each focusing on a particular product. This may tie in with a product team approach linking product development, marketing and manufacturing. The potential capacity of each facility needs to be understood. It should be calculated for different conditions, e.g., normal working, maximum, and break-even. The effects of seasonal demand and long-term trends should be taken into account.

The process span must be chosen. Few manufacturing companies will be involved in all activities from converting basic raw materials to distribution to individual customers. There will be activities that the company does better than others, and it will probably retain these activities. In other areas where the company is not strong and does not have the necessary expertise, it may be appropriate to buy the part or service from a supplier.

The productivity and flexibility aspects of processes have to be investigated in detail. Although the company will be looking for both, it will often find that solutions offer one, but not the other. A good example of this is the transfer line and the batch workshop: the transfer line offers very high productivity but low flexibility; the batch workshop offers very high flexibility but low productivity.

Computer-based simulation of the layout of production facilities is cheap and effective. It allows errors to be found before they are set in concrete. For example: various line configurations can be investigated. Space requirements, flow times, maintenance costs, estimated product lifetimes, and personnel requirements can be identified; annual running costs can be calculated for different line configurations; and management can take decisions based on detailed knowledge of the most important variables.

Human resource issues should be addressed. The right people must be hired, trained, motivated, managed, and kept in the company. The quality issues also must be addressed. Quality is at the heart of customer satisfaction. This applies both to the end customer using the company's product and to internal customers, such as the manufacturing engineer receiving a drawing of an unmakeable part from a designer. Quality is a major issue. It can be approached in many ways—TQM, Quality Assurance, Quality Inspection, SQC (Statistical Quality Control), SPC (Statistical Process Control), and Quality Circles.

Applying computers to areas such as these before the manufacturing strategy has been made clear is not the best solution. Using an example from an area not previously mentioned, it would clearly be wrong to define how IT will be used to gain competitive advantage through the company's relationships with suppliers, until the company's policy towards suppliers has been made clear and communicated in the manufacturing strategy.

Operational managers will set up medium-term production schedules. They will want to see what progress is being made, where bottlenecks are occurring, and what actions can be taken to get back onto schedule. They will also require information on cost and labor used.

Many logistics—related transactions will need to be carried out. They will include purchase orders, manufacturing orders, and shipping orders. Also stock reports will have to be kept highlighting out-of-stock and overstocked parts and products. Actual stock levels will be compared to planned stock levels. Inventory turnover analysis will be made.

If top management lets the introduction of IT within the Production function be carried out solely on the basis of direct productivity gains in well defined activities, they will effectively decrease the possibilities of increasing productivity through inter-functional use of IT. Communications networks are a good example of this. Although a proprietary network may be suitable for the needs of a local area, problems will arise when information has to be transferred from this network to another proprietary network.

Unfortunately, many companies today find themselves immersed in problems resulting from use of many noncommunicating systems from different vendors. Before building interfaces to "overcome" these problems, an attempt should be made to understand how the company can use IT to gain competitive advantage. The resulting requirements for computers in the production function may show that the development of such interfaces is actually a low-priority or unnecessary task.

4.6 FINANCE, ADMINISTRATION, AND HUMAN RESOURCES

Most manufacturing companies have a long history of using computers in the Finance, Administration, and Human Resource functions. Often this was the first, and sometimes the only, function addressed by the EDP Department. Few top managers will feel the need to read this section. After all, they know that their EDP Department has been hard at work for the last twenty years computerizing everything in sight in the Finance, Administration, and Human Resource Department. So, to top managers anxious to move on to the next section, just one question before you turn the page. "Where's the competitive advantage in your systems?"

The following are financial applications that may not be providing competitive advantage:

- General ledger
- Payroll
- Accounts receivable
- Purchasing
- Accounts payable

- Working capital management
- Tax reporting
- Cost accounting
- Sales order processing
- Inventory management
- Fixed-asset management
- Sales analysis
- Budgeting
- Standard costing
- Job costing
- Estimating
- Credit control
- Regulatory reporting
- Cash management
- Treasury management

The history of computing in this area is a story of automating whatever could be automated and replacing headcount expense with system expense. In some cases this may have been the right thing to do. In most cases it will not have been the most relevant. How have these systems improved customer relationships? How have they helped increase revenue? For a function that has such a high responsibility for important resources such as personnel, money, materials and customers, it is surprising that the competitive advantage generated by these systems should be so difficult to perceive. Although these functions appear to be strongly penetrated by IT, and personnel have at least a basic knowledge of using computers it is rare that IT has been used for competitive advantage. Many opportunities exist, and it may be necessary for top management to call in external consultants to take stock of the current situation and make recommendations as to what should be done in the future.

The suitability of the overall architecture of the system needs to be checked and the basic functions of the system must be working perfectly. Controls must be in place to maintain the reliability of information and prevent introduction of corrupt data. The various systems must be closely integrated together, so that information in different systems, such as budget tracking and general ledger, are the same.

Each application should be looked at in detail from the competitive point of view. Does the purchasing system, for example, support the purchasing strategy? If the company is trying to work with a smaller number of suppliers, then the system should be supportive.

Is the cost accounting system relevant? Most cost accounting systems need to be modified to take account of the rapidly decreasing percentage of direct labor in the total product cost. New production philosophies must be reflected in cost accounting systems.

Is there a system in place to make sure that minimum tax is paid? If the company has multinational operations, there may be opportunities to rearrange borrowings to reduce taxation in high tax countries and lessen the effect of double taxation. Is the treasury management system effective in minimizing risks due to financing and operating in a variety of floating currencies? Can the company's lawyers model the potential cost of product liability damages?

Is the sales order processing system providing the type of service customers require? Does the system always select the right discount—in particular for multinational customers with many levels of subsidiaries? Are product configurations checked at order time? Are orders acknowledged immediately? How long do customers have to wait to be informed of delivery dates? In what percentage of cases are these dates not met? What new procedures have been introduced and documented to allow customers to use EDI? When customers telephone in asking for a quick delivery, do they get an immediate answer, or do they have to wait three days while stock levels are checked, prices calculated, and credit validated?

The personnel function will need to maintain applications to support hiring, payment, training, promotion, firing, retirement, and benefits of personnel and their families. Information such as skills required, job descriptions, training specifications, bonus packages, and pay rates will be required. Information will be necessary on hours worked, overtime hours, and vacation taken. Different people will want to make different use of the information. Employees may want to check their payments into health and pension schemes. The Personnel Manager will want to understand the potential exposure to the company resulting from health care and retirement benefits. Production managers may want to correlate turnover rates and job types. Training managers may want to estimate next year's training requirements. Absenteeism may need to be better understood before it can be reduced. Personnel data related to other companies, regions and countries will be kept both to ensure that the company is competitive and to identify new opportunities.

Payroll and pensions, two of the major applications related to personnel, have been computerized for some time. However they have generally been computerized by the EDP Department with little attention being paid to some of the more complex issues that are just as important as the basic functions.

From the employee point of view, payroll and pensions are important applications. The more information that they can get from the systems the better. They will want to know exactly how much tax they will need to pay, to what benefits they are entitled and the cost to them of services the company may provide. If an incentive scheme is in operation, they will want to know their current standing. They will want to know the current and term value of their pension rights, evaluate several possible routes towards retirement, check vacation entitlement, and baby care rights, and their cafeteria benefits. From time to time they may need various forms from the company showing that they are in employment or showing details of their renumeration. Those employees who hold shares in the company or whose earnings are related to production or sales achieved will want to know their current situation. All of these applications, although they are primarily employee-oriented, help the company in the long run. People who know the exact position of their finances will not need to waste working hours trying to figure out exactly how much they can spend before the end of the month.

The sales representative who is at 86% of target at the end of the eleventh month of the sales year will probably make the effort to finish the year on target, whereas a sales representative who does not know the exact position may not bother to make the effort.

Personnel managers will want basic applications, such as payroll and pensions to run efficiently and provide the required service to individual employees. They also need to make sure such systems work effectively from the company's point of view. The system must be able to accommodate full-time employees, those on hourly rates, those working shifts, and so on. Management will require regular reports on the overall situation. Detailed information will have to be exchanged with the costing system along with monthly and annual statistics made available for the company, for individual departments and projects. Absence will need to be recorded and identified as sickness, vacation etc.

In companies where personnel travel a lot, the authorization, control and repayment of travel expenses can be a major task, especially if travel allowances are not the same for all levels and types of personnel. A suitable system can keep the process running smoothly and ensure that cost are kept to a minimum.

The effects of increases due to salary reviews can be simulated in advance of discussion on remuneration so that the overall cost impact for the company is clearly understood. Complex remuneration packages designed to provide high motivation can be modelled and operated. The relationships between remuneration, level and responsibility can be examined. Career paths of existing employees can be examined to see how people evolve in the company. These can be extrapolated to show likely roles in the future. This could show that remuneration costs are going to rise very quickly in the future, in which case remedial action is required. Alternatively they may show that employees are not progressing fast enough up the career ladder, with the result that there will be experience gaps in the future. The effects of losing individual, or teams of, key employees can be examined.

Training requirements will be closely linked to the expected career path. Information will be kept on the content and aims of each training activity, on the training followed by each employee, and on the future requirements. This will enable the human resource function to identify and set up future training activities, and calculate their cost. Increased use will be made of computer-based training, in which the system acts as a one-on-one instructor.

A data base of existing and required skills and experience should be maintained. Some of the open positions will be filled by internal promotion, others will call for recruitment. Information on applicant's skills and experience can also be maintained. This allows rapid matching of their profiles to requirements, therefore, the expected career path and cost of a new recruit can be estimated.

A high quality standard letter can be developed so that applications can be acknowledged rapidly, giving the impression at least that the company is efficient and aware of the needs of potential employees' requirements. During job interviews, potential employees can be shown individually tailored information on how they are expected to grow within the company and on the training that they would receive.

The applications of human resource management and information systems include the following:

- Recruitment
- Payroll

- Pensions
- Benefits counseling and information
- Benefits management
- Training
- Performance appraisal
- Job evaluation
- Job sharing/Flexible working hour management
- Manpower planning
- Absence monitoring
- Human resource administration
- Travel expense management

4.7 INFORMATION TECHNOLOGY

As IT is increasingly used to help gain competitive advantage, the behavior of the IT function will become increasingly important to top management. In the past, IT was considered a passive support function, receiving little management attention.

IT function-related issues for top management include the following:

- The Information resource
- Information Technology strategy
- IT Function performance measurement
- Organization of the Information Technology function
- Facility Management
- Computer Aided Software Engineering
- Knowledge-Based systems
- IT Industry standards
- IT Training

Information is becoming one of the most important resources in the company. In the past, it received little attention, but in the future it will receive as much management attention as other major resources, such as money and people.

Top management must define the IT strategy of the company. The IT function cannot define its strategy independently of the rest of the company. The IT objectives must be directly related to the overall objectives of the company.

Targets must be set for the IT function. These should relate to factors of importance to the whole company, such as reduced lead times, and not to factors internal to the IT function, such as the number of lines of code produced per day. Top management must be involved in selecting the factors and in defining the target values. The targets must be easy to measure and unambiguous.

Organizing the IT function is a major issue for top management. There is a lot of difference between the way that EDP has been organized and run traditionally, and the way that things will have to be done if IT is to be used to gain competitive advantage. The leader of the IT function must have a good business knowledge as well as good knowledge of IT. The subactivities that make up the overall IT activity must be clearly identified and understood. Then it will have to be decided how each one of them will be addressed. It may well be that some of them, perhaps even some of those that appeared most important and were the most time-consuming in the EDP environment should be subcontracted. On the other hand, some subactivities that were neglected in the EDP environment may now appear to be the most important.

Facilities management is an example of this. Companies can subcontract to an independent third party the operation and management of their computer installation. Many EDP Managers would shudder at the thought of this, as it would appear to reduce the size of their empire at one stroke. It does however create opportunities. The extent of contracting-out can vary. Some companies only get the contractor to run the installation on the company site. In other cases, the installation on the company site is scaled down and much of the computing is carried out at the contractor's site. Sometimes, the contractor also takes responsibility for developing new programs as well as maintaining, operating, and managing the current installation. From the point of view of gaining competitive advantage, facilities management often makes sense. Few manufacturing companies today can gain competitive advantage from the way they operate and manage their computer installation. They should use their IT resources for activities that are more likely to help them gain competitive advantage.

Within the IT function, one of the technologies that can lead towards competitive advantage is CASE (Computer Aided Software Engineering). CASE is made up of a set of computer-based tools for planning, developing, and documenting computer systems. It aims to provide a faster, more comprehensive and more accurate way of producing and maintaining information systems. It is not easy to agree to a proper specification for an information system, translate this into separate, independently testable modules, and generate the code. Apart from the very visible activities of analysis, coding, and testing there are other more administrative activities, such as project management and control, quality and configuration management and the production of useful and accurate documentation. The long-term objective of CASE is to automatically translate the business requirements and environment, as specified by the user of computer systems, into a working piece of software. This represents the complete automation of the software development process. At present the complete process can only be automated for very simple applications. However, CASE has already proved invaluable for separate parts of the process. In particular at the design and analysis stages of software development, CASE tools help eliminate errors that would otherwise not be found until much later in the process, at which time their elimination would be expensive in both time and money. Common use of CASE among a group of software developers should lead to improved communication, documentation, project management and configuration management. At the front end of a project, CASE helps in achieving agreement about the precise specification for software, building a conceptual model of the total system and then breaking this down into small manageable independent parts.

The decision by the IT function to standardize on a CASE tool or on other software engineering techniques offers several advantages. People can be trained to be highly productive with the chosen tool, communication between staff can be improved, work can be reused on new projects, software development can be more consistent, and quality control eased.

A technology that can be expected to lead to competitive advantage is that of Knowledge Based Systems (KBS). These are systems that aim to allow the experience and knowledge of humans to be represented and used on a computer in such a way that they increase people's decision-making ability. Knowledge-based systems can be used in a wide range of applications involving research, engineering, planning, configuration, scheduling, analysis, monitoring, control, diagnostic, maintenance, and prediction. They use the same type of process as humans to get to the answer. However they can make use of a much wider range of knowledge and experience, and are not susceptible to emotional influences, fatigue and other human weaknesses. They can provide support in situations where time or money constraints would make it impossible or impractical to rely on human support. For example, the knowledge and experience of plant operators, design staff and line managers can be encapsulated in a KBS so as to be available for plant operators all round the world every minute of the year. In the product development function, a KBS can make available to young engineers the knowledge and experience of their more experienced colleagues. This will help them to respond faster, getting products to market faster and with better quality. In the finance function, a KBS can help an inexperienced clerk take prompt decisions, on for example, a customer's credit worthiness, that would otherwise require referral to more senior staff with concomitant waste of time and lack of service to the customer. In the human resource function, a KBS can be used to check that jobs are rated accurately and consistently, and that remuneration is equitable. In all these cases, people are able to draw on the experience of colleagues who are not present: this is obviously useful in real-life situations but it can also help in the training environment. The knowledge in the system will continue to grow even if some of the human experts leave the company.

There is much discussion as to the policy companies should follow in relation to standards concerning the products of the IT industry. A particular company's policy to standards should result from its IT strategy, and from the way it wants to use IT. If the company wants to be at the very forefront of computer technology it may well have to forego some of the advantages of standardisation. This, though, is a management decision. In the fast growing and fast changing Information Technology industry it is impossible to forecast with any accuracy the products and suppliers that will exist in 5 years time. Standards can only address what has existed for some time. Once a standard has been set, vendors still have to differentiate themselves, and they will have to do this with a product that includes the standard and some other functionality: i.e., is nonstandard. The issues related to standards are close to the technology. They should not be allowed to cloud the search for the best way for the company to use IT for the competitive advantage.

Few companies carry out sufficient IT training. This can be an area where a company can gain a competitive advantage. If a company can actually implement the systems it feels will give it a competitive advantage, it will already have gained an advantage over all those companies still struggling to bring projects to successful completion. Training is required for the IT professionals so that they can operate systems without great expense, and develop new systems quickly. It should not be forgotten that the computer can play a major role in training IT professionals. The users of IT should also receive IT training. They need to understand why IT is being used, how they can make best use of it, how it affects their job, and how they can use it to improve the company's competitive stance.

Training offers an opportunity to gain knowledge and skills. It can also indirectly support the company culture and help change the behaviour of trainees. Training company personnel to support use of IT as a means of gaining competitive advantage is an important issue for top management.

4.8 CUSTOMERS, SUPPLIERS, AND COMPETITORS

The previous sections of this chapter have pointed to some of the many activities in a manufacturing company which may provide an opportunity for using IT to gain a competitive advantage. It is not enough though, to only look within the company for such opportunities. There are also many opportunities outside the company, for example in the relationship with customers and suppliers.

The following are some of the areas where IT can be used to gain competitive advantage in relationships with customers, suppliers, and competitors:

- Electronic Data Interchange (EDI)
- Electronic Point Of Sales (EPOS)
- Electronic Shop Window (EPOS)
- Telemarketing
- Terminals on customer sites
- External Data Bases
- Competitive Intelligence
- Common systems
- After sales service
- Telemaintenance

EDI is the exchange across a telecommunications network of documents and other routine paper information such as receipt advices, purchase orders and advance shipping notices between the computer in one company and the computer in another company. The information is transmitted in a standard format in which the various contents of the documents are arranged in a predescribed way. EDI is already heavily used in industries such as automotive, chemical, and retail. Companies in these industries which are not using EDI risk putting themselves at a competitive disadvantage. EDI helps reduce errors by eliminating re-entry of data. Time cycles are greatly reduced. Whereas by post it may take 3 or 4 days for information to reach a supplier, with EDI it only takes a few minutes. Many companies also claim that use of EDI reduces the cost of processing an order by up to 75%. Given the advantages of improved quality, and reduced time and cost it is easy to see why EDI is becoming so widespread.

As with any technology there are companies that will gain greatly from use of EDI, and others that will only achieve minimum benefits. To make competitive use of EDI, companies will have to do more than simply transfer purchase documents electronically. They will have to use this basic ability as leverage for making other changes. They will have to do more than just cut down on postage and clerical costs. To do this they will have to make organisational and work flow changes to take advantage of the potential offered by this new technology.

Initially, EDI primarily addressed documents such as purchase orders, which do not contain a lot of information, but are very numerous. Now, EDI is beginning to be used to transfer engineering documents. These may come from a CAD system and they tend to contain a lot of information and be less numerous. Again the gains are clear. When a paper drawing is sent by post it may take several days to arrive at a supplier's site. When it is sent electronically it may only require some minutes. This opens up possibilities for much closer cooperation between companies and their suppliers at the product development stage. There can be a real interactive partnership between engineers in the two companies. This can create a competitive advantage for the supplier relative to other suppliers not cooperating in this way.

Electronic Point Of Sales (EPOS) is another area that offers companies the possibility to gain competitive advantage. Beyond the advantages of fast-moving check-out lines, a detailed receipt and the certainty that one really has received '10 cents off', is a panapoly of potential benefits. Data from the cash registers can be used in the store for stock control, and in the Marketing Department for improving product offerings. In the store, it can be used for automatic stock-reordering, up-to-date stock control and detailed reporting of sales. In just a few days, Marketing can receive information for which it previously had to wait months. The effect of promotions, pricing, different displays, new products can be seen much quicker, and prompt decisions can be taken to maximise sales. EPOS has spread rapidly in supermarkets, gas stations, and department stores. There are many other sectors where it could be applied.

Information Technology lies at the heart of the Electronic Shop Window (ESW). The information on company's products can be put into a database that can be accessed by its customers. Customers no longer need to leaf through out-of-date catalogues or wait for a salesman to call. They can call up the database over a communications network or look up information on a diskette regularly updated by a manufacturer. Other applications can be built up around the basic services so that customers can see the status of their orders and check the position of their accounts. Instead of only waiting for potential customers to look in the Electronic Shop Window, the company can also identify potential customers and send them an electronic mail-shot or a fax giving them details of a new product or service.

Telemarketing offers companies a new way to increase their sales activities without taking on large numbers of new sales representatives. Companies can access a geographically wider set of customers and contact potential customers who would not be suitable targets under standard selling conditions.

Installing sales terminals in the customer purchasing office is an effective way of increasing business and causing problems for competitors. Few companies will want to have terminals from several companies in their purchasing offices so there is a great advantage in being first.

External databases are a source of information that many companies can benefit from. Usually the information is available in paper format as well, but to have the same volume of information available on paper would not be possible. There is also the time element. Information can be obtained rapidly from a database. If it has to be first located in paper form, then purchased, then transported to the user, much time will be lost. There is a vast range of information available now on databases. It includes data at the continent, country, region, state, city, town, and street levels. There is information on government regulations. There is information on sales figures. There is information on industrial production and on individual products.

IT can be used to help increase the information that a company holds on its competitors. One way to do this is to equip the sales force (or even specialists in this area) with portable computers. As they visit customers, they can input data on competitors prices and promotions, information on new products announced by competitors, data on stocks of competitor's products and competitor's price lists. This information can be supplemented with competitors annual reports, newspaper articles, market studies, and competitor's product literature and catalogues.

Suggesting to a company's suppliers that they make use of the same computer systems can create a strong bond. There are good reasons for using the same system, for example the use of common data formats eliminates the need for data translation between the formats of different systems. One example of this is in the exchange of CAD data between automotive manufacturers and their suppliers. If a manufacturer and a supplier use different systems there is the danger that errors will be introduced and time lost during the transfer of data between the two systems. If there is a common system the exchange process should be rapid and error-free.

Many large companies now make available information on availability and supply of spare parts over a network. Companies in consumer sectors such as automotive and white goods have completely reorganised their parts organisations to take advantage of the use of telecommunications and computers to handle thousands of distributors spread over large areas. Similar adjustments to siting of production facilities may be possible.

Telemaintenance allows companies to provide assistance to customers with problems without physically visiting the customer site. Diagnostics are run to verify the problem and solutions are proposed to the customer. If these fail it may be necessary to send a new part to the customer. If this does not solve the problem or the customer is unable to carry out the maintenance activity then a maintenance engineer can be sent to the customer site. In most cases, the problem is resolved very quickly and very cheaply. Computerised call handling and tracking informs engineers and managers alike of problems, solutions and work patterns. Analysis programs can help identify recurrent problems, eventually leading perhaps to product modifications. The sooner a problem has been identified, the sooner it can be solved. Equipment can be serviced quickly and put back into production as soon as possible.

This part of the book has demonstrated that there is a very large number of potential applications of IT for a typical manufacturing company. Without a well-defined strategy, applications will be implemented at random, and the result may show no increase in productivity, no improvement in market position and no improvement in the way the company works. On the other hand, a large investment in IT will be apparent, and top management will naturally question the wisdom of having made, and continuing to make such an investment.

Rather than taking a random approach to IT, companies should use it to support their strategy to gain competitive advantage. This implies first understanding how they will gain competitive advantage, and then identifying and implementing the relevant IT systems.

BIBLIOGRAPHY

Peter F. Drucker. *Innovation and Entrepreneurship.* Heinemann, London, 1973.

Peter F. Drucker. *Management.* Pan Books Ltd., London, 1981.

Eric Gerelle and John Stark. *Integrated Manufacturing: Strategy, Planning and Implementation.* New York: McGraw-Hill, 1988.

Joseph Harrington, Jr. *Understanding the Manufacturing Process.* New York: Marcel Dekker Inc., 1984.

Joseph Harrington, Jr. *Computer Integrated Manufacturing.* New York: Marcel Dekker Inc, 1986.

Sir Basil H. Liddell Hart. *Strategy.* New York: Frederick A. Praeger, 1968.

Richard Lubben. *Just-In-Time Manufacturing.* New York: McGraw-Hill, 1988.

Michael Porter. *Competitive Strategy: Techniques for Analyzing Industries and Competitors.* New York: Free Press, 1980.

Michael Porter. *Competitive Advantage.* New York: Free Press, 1985.

Richard J. Schonberger. *World Class Manufacturing: Implementing JIT and TQC.* New York: Free Press, 1987.

Shigeo Shingo. *A Revolution in Manufacturing: The SMED System.* Cambridge, MA: Productivity Press, 1985.

John Stark. *CADCAM Management—Implementation, Organization and Integration.* New York: McGraw-Hill, 1988.

John Stark. *Handbook of Manufacturing Automation and Integration.* New York: Auerbach, 1989.

Robert E. Umbaugh, ed. *Handbook of MIS Management,* 2d ed. New York: Auerbach, 1988.

INDEX

INDEX